Dublin's Meath Hospital
1753–1996

Dublin's Meath Hospital
1753–1996

Peter Gatenby

TOWN HOUSE
DUBLIN

Published in 1996 by
Town House and Country House
Trinity House
Ranelagh, Dublin 6
Ireland

Text copyright © Peter Gatenby, 1996

All rights reserved. No part of this publication may be reproduced, stored in a retrieval system, or transmitted in any form or by any means, electronic, mechanical, photocopying, recording or otherwise, without prior permission in writing from the publishers.

British Library Cataloguing in Publication Data. A catalogue record for this book is available from the British Library.

ISBN 1–86059–040–3

Cover illustration: a painting by W G Spencer showing a modern view of the Meath Hospital in the foreground and the new genito-urinary unit in the background (*c.* 1978).

Typesetting by Red Barn Publishing, Skeagh, Skibbereen
Printed in Ireland by Betaprint

Contents

Acknowledgements	vii
Introduction	1
Chapter One Early Days in the Coombe, 1753–1821	3
Chapter Two The New Hospital, Graves and Stokes, 1822–5	20
Chapter Three Fevers and Famine, 1826–50	29
Chapter Four The Smyly Appointment Controversy, 1861	40
Chapter Five Hospital Conditions and Staff, 1860–87	45
Chapter Six The Meath Hospital in the Confident Nineties, 1890–9	54
Chapter Seven Queen Victoria's Visit and the Appointments Controversy, 1900–12	78
Chapter Eight Infectious Diseases, 1894–1914	89
Chapter Nine War Years, 1914–19	93
Chapter Ten Towards Financial Crisis, 1920–30	97
Chapter Eleven The Dormant Thirties, 1930–9	105

Chapter Twelve
The Emergency Years and Tuberculosis, 1939–48 — 112

Chapter Thirteen
The Advance of the Knights, 1949–50 — 119

Chapter Fourteen
Back to Normal, 1951–2 — 130

Chapter Fifteen
The New Genito-Urinary Unit, 1953–60 — 136

Chapter Sixteen
Medical Specialisation, 1961–72 — 158

Chapter Seventeen
Further Specialisation, 1972–92 — 163

Chapter Eighteen
Nursing, Physiotherapy and Social Work, 1807–1996 — 170

Chapter Nineteen
Social Occasions and Lighter Moments — 181

Chapter Twenty
Some Reflections — 185

Chapter Twenty-One
Towards Tallaght — 189

Notes — 192

Appendix I
Act of Parliament George III, Regis, Cap LXXXI — 196

Appendix II
Officers and Staff of the Meath Hospital
and County Dublin Infirmary — 198

Index — 203

ACKNOWLEDGEMENTS

I first wish to thank the publication committee of the Meath Hospital for entrusting me with the task of writing a history of the hospital.

Mairead Shields, personnel manager, facilitated me in providing a room where I could study all the minute books and records of the hospital. Dr Brian Mayne had done the pioneer work already in listing all the available sources of historical information which were lying in dusty shelves and cupboards.

I am very grateful to Professor Davis Coakley, dean of the faculty of Health Sciences of Trinity College, who has given me valuable advice, information and encouragement.

The following medical colleagues read sections of the manuscript and corrected mistakes and advised me: Brian Mayne, Brandon Stephens, David Lane, Derek Robinson, John Barragry, Dermot O'Flynn, Ian Temperley, Brian Keogh and Gerry Hurley.

I wish to thank all of them, and also the many people who gave me valuable information. These include Michael Pegum, Desmond Darling, Barbara Stokes, Joe Galvin, John Mackey, Desmond O'Neill, Fergus O'Ferrall, Mrs Daphne Wormell, Mrs Henry Boylan, John Fleetwood and C Garrett Walker.

I wish to mention especially the assistance of Ms Margot Hanly, former assistant secretary to the hospital, whose memory goes back a long way and who maintained valuable records in a fascinating scrapbook.

In writing the history of the nursing school I obtained great help from Kathleen Brennan, Margaret McCarthy, Evelyn Doherty, Mary Lennon, Eileen Sheridan, Mary Cotter, Elizabeth and Angela Hoey.

The photographs were prepared skilfully by David Smyth of the TCD faculty of Health Sciences and by Tom Walsh of the Meath Hospital.

Last but not least I must record my grateful thanks to Gay Williams for her special expertise in not only doing all the secretarial work but also in advising me on the presentation of the material.

INTRODUCTION

Since its foundation in 1753, the Meath and County Dublin Infirmary has had a very distinguished history, especially in the nineteenth century when it was the leading hospital during the period of Dublin's international medical fame. 'The Meath' is due to end its separate existence in 1997, when it will merge with the Adelaide Hospital and the Children's Hospital, Harcourt Street, to form the new hospital at Tallaght in south-west Dublin. The governors and the staff have persuaded me to take on the task of writing an updated history of the hospital. I have done this mainly by extracting information from the minute books of the governing board (previously known as the 'standing committee' or 'joint committee') and the medical board. These are available from the late 1700s.

Sir Lambert Ormsby published his *Medical History of the Meath Hospital* in 1882, and a second edition appeared in 1892. His history is dominated by fascinating biographies of the medical staff in the Victorian period. I began by writing the history from the time that Ormsby concluded his account in 1892, but I soon found that it was very difficult to start the story in the middle, without referring to significant events and legal enactments of the past. So I subsequently decided to start at the beginning and write a history from the foundation of the hospital.

Though I served the hospital as physician for seventeen years from 1957, I regret that I was not aware then of the fascinating details of its history, which is studded with medical and nursing personalities. John Cheyne, Robert

Graves and William Stokes were all on the staff of the hospital during the golden period of Irish medicine. Their names have become immortalised throughout the medical world because of their writings, which give original classical descriptions of diseases. I have made reference to these and other distinguished medical men, where their opinions and work are mentioned in the original records of the hospital.

From humble and squalid accommodation in the Coombe, the hospital moved to its present site in 1822. There it gradually expanded, as money became available through donations or bequests. The hospital played a major role in coping with the fever epidemics before, during and after the famine years. This book is essentially the story of the conditions in the hospital for the staff and patients over 244 years, reflecting the prevailing political and financial situation. The history recounts the tension between the governing board and the medical board, because the latter had been granted the legal right, by ancient Acts of Parliament, to appoint their successors.

A major upheaval occurred in 1949, when members of the Catholic society, the Knights of St Columbanus, took over the hospital. Fortunately, an all-party parliamentary action resolved this crisis in 1951 and the hospital resumed normality. After this, the most outstanding event of the twentieth century for the hospital occurred when the remarkable T J D (Tom) Lane developed the specialised genito-urinary department in 1955, bringing international fame to the hospital. Since then, specialisation in other departments of the hospital has steadily increased, in spite of the severe pressure of accident and emergency work and overcrowded conditions.

The hospital staff are now looking forward to moving to modern accommodation at Tallaght and combining with the Adelaide and Harcourt Street Children's Hospital, and to continuing together their great and long traditions of service to the sick.

CHAPTER ONE

Early Days in the Coombe, 1753–1821

The Meath Hospital was first opened in the Upper Coombe in rented accommodation in 1753. Then, in 1756, it moved to Skinner's Alley. After four years, in 1760, it moved to Meath Street, and from there to Earl Street in 1764. Eventually a site was obtained in the Coombe and the foundation stone of the new hospital was laid by Lord Brabazon on 10 October 1770. The hospital was to remain there for fifty years. This area was in the Earl of Meath's 'Liberty', so-called because it was outside the jurisdiction of the Lord Mayor, and it lay along the 'Coombe', or valley, of the River Poddle, a tributary of the Liffey. Although nowadays not much better than an underground sewer, this river was once navigable, filling the moat of Dublin Castle and working the ancient mills near Dame Street. Before becoming the Liberty, these lands had belonged to the Abbey of St Thomas, a twelfth-century foundation, which stood in Thomas Street. The abbey was granted by Henry VIII to William Brabazon, ancestor of the Earls of Meath, for ever, as a reward for military service, at a rent of 18s 6d per annum.[1]

The Population Act of 1814 estimated the population of Dublin to be 175,319, a figure larger than any city in England except London.[2] Great buildings had been constructed in Dublin, such as the frontage of Trinity College, the Custom House, Four Courts and the King's Inns. The Irish Parliament was well established at College Green, yet despite being the second city of the Empire, distinguished by many magnificent buildings and wealthy inhabitants, the poor lived in dreadful conditions. Constantia Maxwell quotes horrifying contemporary descriptions of the filth and squalor of the Dublin poor.[3] 'Poverty, disease and wretchedness exist in every great town,' wrote Curwen, an Englishman, 'but in Dublin the misery is indescribable.'

The Reverend James Whitelaw, rector of St Catherine's Church in Thomas Street and a governor of the Meath Hospital, described the Liberties where the Meath Hospital was founded as follows:

> The streets are generally narrow: the houses crowded together: the rears or backyards of very small extent, and some without any accommodation of any kind. Of these streets, a few are the residence of the upper class of shopkeepers or others engaged in trade: but a far greater proportion of them, with their numerous lanes and alleys, are occupied by working manufacturers, by petty shopkeepers, the labouring poor, and beggars, crowded together to a degree distressing to humanity.
>
> I have frequently surprised from ten to sixteen persons of all ages and sexes, in a room not fifteen foot square, stretched without any covering, save the wretched rags that constituted their wearing apparel.
>
> This crowded population wherever it obtains is almost universally accompanied by a very serious evil—a degree of filth and stench inconceivable. Into the backyard of each house, frequently not 10 feet deep, is flung from the windows of each apartment, the ordure and other filth of its inhabitants; from which it is so seldom removed, that I have seen it nearly on a level with the windows of the first floor.[4]

It is usually recorded in various accounts of the Meath Hospital that the first notice of the hospital appeared in the *Gentleman's and Citizen's Almanac*, published by bookseller John Watson in 1754, as follows:

> The Meath Hospital in the Coombe was opened on March 2, 1753, supported hitherto by a benefit play, some benefactions, and annual subscriptions of several of the principal inhabitants of the Earl of Meath's Liberty, and other well-disposed persons who judged that an institution of this nature was much wanted in a part of the town remote from city hospitals and greatly thronged with the industrious poor. Messrs Alexander Cunningham, Redmond Boat, David McBride, and Henry Hawkshaw, surgeons, attend daily in their turns, and all serve without fee or reward.

At that time the other hospitals in Dublin were the Charitable Infirmary (founded 1718) at Inns Quay (and later in Jervis Street), Steevens' Hospital

(1720) and Mercer's (1734). The three specialised hospitals were the Rotunda, or Lying-In (1745), St Patrick's Hospital (1746) for mental disease, and the Hospital for the Incurables (1744). As Boxwell writes: 'they were naturally all Protestant foundations, for they were inaugurated in the worst period of the Protestant Ascendancy; and while the Protestants in Ireland then were in a minority of one to six through the action of the Penal Laws and land confiscation, they were the only people who had any money and the will to spend it on this type of charity'.[5]

The Penal Laws were repealed in 1778 and 1782 and the outburst of philanthropic activity in the late eighteenth century in Dublin included joint support from the well-to-do Catholics, such as that evidenced by the foundation in 1790 of the Sick and Indigent Roomkeepers' Society in Dublin, which was and still is strictly a non-denominational charity.[6]

Up until 1700 there was considerable prosperity amongst the woollen weavers in the Coombe area. The industry flourished because of a good supply of raw material and the cheapness of labour, and a favourable export trade had developed. However, as a result of petitions by clothiers in England, restrictions on the Irish trade were instituted by the English Parliament and the industry was destroyed, leading to great poverty in the area. The Meath Hospital was founded there to provide succour to the impoverished local population.[7]

During the bicentenary of the hospital in 1953, Professor Widdess referred to an advertisement 'recently discovered' by Miss Mary Ennis which had appeared in *Faulkner's Weekly Newsletter* in March 1753 and which undoubtedly refers to the establishment of the Meath. It informed the public:

> that the Surgeons belonging to the Charitable Surgery in Spring Gardens, Dame Street, finding that the greatest number of people to apply to them came from the Earl of Meath's Liberty and parts adjacent, and being convinced that the design of relieving the distressed poor will be much better answered if the place of attendance is changed to the part of the town which is very distant from the Hospitals, they have therefore for the convenience of them taken a house on the Upper Coombe near Meath Street, which was opened on March 2, where they propose to attend to dress, distribute medicines and give advice from 8 o'clock every morning until 10. N.B. Any person willing to encourage the above charity will please give their benefaction to Mr Philip Crampton of Crampton Court, Joseph Terry of Braithwaite Street and Gerald Bryen, opposite the Weavers' Hall on the Coombe.

Of the original surgeons, David McBride and Redmond Boat soon resigned, but Alexander Cunningham and Henry Hawkshaw continued in office until they died, after nearly thirty years of service. They lived to move into the new hospital building in the Coombe which was opened in 1773. The original two physicians appointed in 1754, Thomas Brooke and Francis Hutchinson, resigned within a year.

The 1774 Act
When the new building in the Coombe was opened in March 1773, the hospital board presented a petition to the Irish House of Commons asking that the hospital be constituted as the County Infirmary of Dublin. The petition stated that, from the first, the physicians and surgeons had served without fee or reward and had given a considerable sum of money from their own pockets to procure or erect hospital buildings. The petition also declared that the annual £100 salary to the County Surgeon would be given to the hospital funds. In exchange for this surrender of salary, the medical board was subsequently granted, by an Act passed in 1774, the legal right to appoint medical staff when vacancies occurred.

The 1774 Act stated that:

> the present Physicians and Chirurgeons of said Meath-Hospital shall be appointed Physicians and Chirurgeons of said intended Infirmary, for the County of Dublin: and that it may be lawful for said Physicians and Chirurgeons, or a Majority of them, to elect a Physician or Chirurgeon in the Room of any Physician or Chirurgeon who from Time to Time, by Death, Removal, or otherwise shall make a Vacancy in said Hospital, in Consideration of their having served said Hospital gratis these seventeen years past, and their having had a principal Share in the Support thereof, during that period, and in erecting said Building, as well as the relinquishing, in Behalf of themselves and their Successors in said Hospital, all Claim or Title to the Annual Salary of One Hundred Pounds, which they otherwise would be intitled to in Consequence of said Act.

This Act clearly conferred on the medical staff the privilege of appointing their successors. As will be seen, this concession was to give rise to problems in the future. Since 1774 the official title of the hospital has been designated 'The Meath Hospital and County Dublin Infirmary'.

1807–13

Little is known about the governors and medical staff in the early years of the Meath Hospital. However, the minute books which were kept from 1807, over 20 years after its accommodation in the new building in the Coombe, provide us with some interesting details.

At the governors meeting on Monday, 6 April 1807, the attendance was recorded [8] as follows:

Edward Allen in the Chair
Saml. Bewley
Arthur Guinness
Geo. Maquay
Thos. Crosthwaite
Benj. Guinness
Pat. Roney
Josh. Hone
John Barrington
Solm. Richards
Peter La Touche Jnr.
Josh. Goff
Chrs. Humphrys
Edw. Swanwick
John Leland Maquay
Rich. Verschoyle

The two medical members present were the surgeons Patrick Cusack Roney and Solomon Richards. The names of the lay members show some well-known Dublin family names, such as Bewley and La Touche. The first Arthur Guinness (1725–1803), the founder of the famous brewery, was treasurer of the hospital from 1775 to 1797. Benjamin Guinness was his brother.[9]

The 'Establishment' of the hospital was recorded as follows in the same minutes:

Francis Barker, Physician.
Thomas Egan, Physician.
Patrick Roney, Surgeon.
Solomon Richards, Surgeon.
Richard Dease, Surgeon.
Cusack Roney, Surgeon.
James Bolton, Register and Collector @ £40 p.a.
James Dunn, Apothecary @ £50 p.a. with apts. coals and candles.
Anne Going, Housekeeper @ £30 p.a. with apts. coals and candles.
Anne Hoey/Anne Kavanagh, Nurses 6 guineas p.a. and 2 guineas for tea and £1 for beer.
Margaret Keogh, Servant 4 guineas p.a. 2 guineas for tea and £1 18s 0d for beer.
Bernard McDonough, Porter @ 3 pence per day, diet and lodging and one suit of clothes per year.

Ormsby gives biographical sketches of the medical staff of this period.[10] Francis Barker was elected Professor of Chemistry at Trinity College after a few years as physician at the hospital. Together with John Cheyne, he wrote on typhus fever epidemics. Thomas Egan, from a Galway family, was also physician to Cork Street Fever Hospital. Patrick and Cusack Roney were father and son. Patrick had another son, Thomas, who was also to be appointed surgeon to the hospital. Cusack was the most distinguished of the family and he was elected president of the College of Surgeons in 1814 and 1824. Solomon Richards was president of the College of Surgeons on four occasions (1794, 1803, 1808 and 1818). He had a large and lucrative practice and was said to be the fattest surgeon in the United Kingdom and only able to descend from his carriage sideways. Richard Dease, whose father William had also been surgeon to the Meath, died prematurely from septicaemia acquired from an anatomical dissection. This family will be referred to later in relation to the Dease Memorial.

Further extracts from the minutes give some idea of the conditions in the hospital in 1807.[11] A committee was appointed to enquire into the internal regulation of the hospital and reported on 20 April 1807 as follows:

> With respect to the patients they find there is not either Clothing or Washing provided for them, in consequence several are without any body linen, and those who have of their friends don't wash for them and it is never done.
>
> The Beds have a dirty appearance and the Quilts are the common Horse Clothes which cannot be washed. The Blankets are mostly good and are washed or scoured almost once in 6 or 9 months. The sheets are of Grey Linen and from the colour look dirty. The Bed Ticks are of coarse Grey Sacking and the straw is changed about once in three months. The hours for meals are Breakfast 8 o'clock, Dinner 2 o'clock and Supper 7 o'clock, but they are not attended with regularity.
>
> There is scarcely any convenience for serving the Provisions to the Patients. The Bread is carried in the Nurses' Aprons and the meat for Dinner is put into the Porringer in which the Broth is and there is neither Plates, Dishes, Knives, Forks or Spoons provided. There is not any allowance of milk for whey in consequence the Friends of the Patients frequently bring them milk.
>
> The Bread and Meat we found good. The milk was poor. The housekeeper informed us that the wards are scoured twice a week and under the beds mopped every day. There is not any regular way for distributing coals, the quantity to be given and the Fires to be lighted are

entirely under the directions of the Housekeeper.

She looks upon it as part of her duty frequently to go through the wards and to have the Candles put out and everyone in bed by 10 o'clock.

There is only two nurses allowed to the Hospital and in cases of severe illness one of the patient's Friends sits up all night.

There is scarcely any conveniences or accommodation for the Nurses or Servant.

The Apothecary keeps a Pass Book with the Druggist and when any Medicines is wanting, he sends an order for it without directions from the Medical Officers. He looks upon it as his duty frequently to visit the Wards.

There does not appear any regular form of Recommendation for the admission of Patients to be signed by the Governor who recommends.

In case of death the Body is not removed from the ward but remains there until taken away by the friends of the deceased or buried at the expense of the Hospital.

The committee further adds that every person applying for medical assistance is attended to without any recommendation and there is not any Registry kept of their names or residencies.

The following week, on 27 April 1807, the governors resolved that:

> Body Linen should be provided for the patients and that the said committee do immediately get made, 50 shirts and 50 pocket handkerchiefs, 50 shirts and 50 nightcaps. And that they be also empowered to make necessary arrangements for washing and increased quantity of clothes.

In May 1807, the standing committee asked the medical officers to give their views on the 'admissibility of patients'. On 12 June 1807, Thomas Egan, MD, replied on behalf of the medical board:

> It is generally been the custom in this and all similar Infirmaries to receive patients in all disorders deemed curable if properly recommended, but here persons suffering under disorders requiring immediate assistance to preserve life and those meeting with severe accidents are instantly admitted without any recommendation. But since the Establishment of the Lock Hospital, it has not been deemed expedient to receive venereal cases except very few under very peculiar circumstances, and contagious disorders have been lately entirely excluded because there is the Fever

Hospital in this neighbourhood and the bad construction of the house was found to render them highly dangerous to the other patients.

The Board are of the opinion that the admission of cataracts and sore legs (in many cases of which they can be eminently useful) is highly proper, but they will carefully attend to the suggestions of the Committee so as to render the institution as extensively useful as possible with the smallest expense, by taking in such cataracts only as are fit for immediate and expeditious operations, and by discharging from the house sore legs and other disorders which after a reasonable trial shall appear more proper objects for other charitable institutions.

The standing committee again enquired into the work of the medical officers in 1809 and received the following from the medical board on 17 April:

The Physicians attend alternately each for two months twice a week, viz. on Mondays and Fridays for Externs, and as often as they deem it necessary for Interns.

The Surgeons attend daily each taking the care of the Intern and Extern Surgical patients for one month in rotation; on the Surgeon of the Month rests the entire responsibility for that time; by him all accidents are attended and all recommendations received, and who in case of Sickness or extraordinary emergency of business must provide one of the other gentlemen to attend for him. Each Surgeon continues to attend the patients admitted or operated on by him, but in case any casualty should cause his absence, the surgeon of the month attends the patient; by this arrangement the house has the advantage of the attendance of two, three or more Surgeons daily.[12]

Though these enquiries from the standing committee to the medical staff showed that it was exercising some normal authority over the medical board, this did not apply to medical appointments. The medical board practised their legal right, already referred to, in appointing their successors. For example, in May 1809 it is recorded in a simple bald statement that 'the medical officers communicated through Surgeon Richards to the Committee that Dr Barker had resigned and that they had appointed Dr Todderick in his place as Physician to this Hospital, and that they had also elected Thos. Hewson Esq as Surgeon to the Hospital'. Again, on 18 November 1811, it is recorded that 'the Medical Governors communicated to the Committee that they have elected Doctor John Cheyne (Licentiate of the King's and Queen's College of Physicians of Ireland) to this Institution in the room of Dr Todderick resigned'.

John Cheyne[13] was a Scotsman who had been at the battle of Vinegar Hill as a government army medical officer during the 1798 rebellion. He decided to stay in Dublin to practise medicine and was appointed Professor of Medicine at the College of Surgeons in 1813. He was surgeon to the Meath for only five years and was then appointed to the House of Industry Hospitals in 1817. He was a prolific writer and a major influence in editing the excellent Dublin Hospital Reports. In 1812 he published *Cases of Apoplexy and Lethargy with Observations on Comatose Diseases*. His name has become immortalised by his description of the ominous waxing and waning of respiration seen in terminal illness, and known universally as Cheyne–Stokes respiration. William Stokes described this phenomenon again in 1846.

Appeals for Financial Support, 1811
There was a considerable difficulty in raising funds to keep the hospital going. Various plans for extending the premises to meet the increasing number of patients were considered but not followed up because of lack of money. In 1811, the standing committee applied to the Chancellor of the Exchequer, W W Pole, for an annual parliamentary grant. He replied that, as the hospital was a County Infirmary, it was not eligible for such a grant, but that the Road Act and other Acts indicated that the Grand Jury of the county of Dublin was legally responsible for granting funds for improvement of the building of the County of Dublin Infirmary.

In 1812 an extensive 'memorial' was presented to the Grand Jury of the county of Dublin, giving the full history of the development of the hospital and describing its urgent needs, and requesting meaningful increased financial support. A similar memorial was sent simultaneously to the Chief Justice and the other judges of the King's Bench, requesting them to 'recommend the subject to the Grand Jury'.

This memorial showed that the annual income of the hospital was as follows:

County Presentment	£600
Parliamentary Grant	gross £100
Interest upon £400 5% Govt. Stock & £200 lent on mortgage	£32
Annual Subscriptions	about £170
Annual Charity Sermon & Donations	about £260
TOTAL 'Certain v Uncertain' Ann. Income	£1,162

> The Committee have still to regret that the Infirmary of the first County in Ireland is totally inadequate to the wants of its extensive and increasing population, the present Building being not only insufficient for the Number of Patients requiring admission but totally defective in the needful accommodation even upon the present limited scale of the hospital.
>
> These deficiencies principally affect the Medical and Surgical Departments and are fully stated in a report made by the Medical Officers and annexed hereto. In addition to which the Committee have to state that there is no Apartment either for the Housekeeper or Register, the former being obliged to inhabit the Board Room and the latter to reside out of the House.[14]

The report of the medical board gives a vivid description of the hospital in its Coombe premises at that time.

> The Public Hall (the only place at present for the reception of Extern patients) is dark, ill-ventilated and small. Hence it is difficult to preserve any order among the Extern Patients, who are often so crowded together as to block up the entrance of the Hospital.
>
> There is no Apothecaries Shop. A portion of the Hall, in front of the window, is railed in for dispensing medicines—and as the extern patients are always supplied with the Medicines prescribed before they leave the hospital, the arrangement in the Hall seems to give rise to much of the Crowd and Clamour.
>
> Accidents and Sores are dressed in the Lobby of the staircase, which has no light but what is derived from the sky-light in the roof; yet this is the only part left for performing the Minor Surgical operations; and as being in the thoroughfare between the Hall and the rest of the House, the Dressers are liable to constant interruption and the patients to considerable risk.
>
> The room for the reception of extern patients has also been made an Apothecaries Store. In this Apartment, which is also small, imperfectly lighted and incapable of proper ventilation, the Medical Officers seldom prescribe for less than one hundred patients daily; many of these capable of communicating contagion. The duty of the Hospital is thus a service of great danger, and the want of proper accommodation for the reception of patients infallibly contributes to the spreading of various contagious diseases, particularly fevers, and the contagious diseases of children.
>
> There is no consulting room for the Physicians and Surgeons.
>
> In the present state of the hospital no attention whatever can be paid

to the comfort of the pupils. Yet it can scarce be questioned that this education is a matter of public and general concern. From the want of an Operation Room, the greater surgical operations are performed in the open wards, in the midst of the sick, much to the inconvenience of the surgeons and pupils, to the distress of the person operated upon, and to the annoyance of all patients who cannot be removed.

Two or three small rooms apart from the noise and bustle of the House are most desirable in every hospital, for those patients who must submit to capital operations, or who have sustained severe accidents.

A proper apartment is also greatly wanted to the reception of the bodies of patients who die in the hospital and for the anatomical investigations which are required by the Coroner.

The hospital is upon a scale totally inadequate to the number of applications for admission. The physicians and surgeons are daily obliged to send away persons in the most pressing need of medical and surgical assistance, many of whom it is feared perish in neglect, and when accidents are numerous, it often becomes necessary to put two patients in one bed, a circumstance disgraceful to a regular hospital.

This description of overcrowding of out-patient facilities is familiar to those of us who have long memories or who have witnessed conditions in Third-World countries. However, the spectacle of major operations performed in open wards in front of other patients must have been horrifying, especially as there was no anaesthesia and many of the operations at that time would have been amputations.

In spite of the lengthy and detailed application to the Grand Jury of the county of Dublin, no increase in the annual grant of £600 was made, so the governors instituted a general appeal to the public for financial support. In 1813, eight hundred copies of a circular letter were distributed to the 'Principal Inhabitants of the County of Dublin'. This described the serious defects in the hospital, and explained that the hospital was £279 in debt, that the certain income was about £730 and that the expenditure was about £1,500.

Copies of letters to the Duke of Leinster and Lord Powerscourt ordered by the standing committee are in the minutes of the meeting of 15 February 1813.

To His Grace the Duke of Leinster

My Lord,
I am directed by the Committee of the Meath Hospital and County Dublin Infirmary to inform your Grace that the extensive and most valuable

charity had long been patronized by your Grace's late father, who was also a liberal benefactor to its funds having been a subscriber of 5 guineas a year which subscription has remained unpaid since the year 1799.

The Committee are convinced that it is only necessary to make your Grace acquainted with the circumstances of this institution to insure your Patronage and Assistance to it, and therefore have desired me to enclose for your Grace's perusal a copy of the Annual Report for 1812. I shall call in a few days for the purpose of receiving any commands your Grace may have to communicate.

A similar letter was drafted to be sent to Lord Powerscourt reminding him that his father's usual annual subscription of 10 guineas had not been paid since 1809. No record has been found of a response to these letters.

Donation by Thomas Pleasants, 1814

This appeal resulted in a single highly significant donation of £6,000 in 1814, which was a landmark in the history of the hospital. This came from Mr Thomas Pleasants of William Street. The minutes of the general meeting of the governors on 4 April 1814 record a statement from him accompanying his donation which contains the following comments:

As he supposes, being a thing usual, that this grant, a grant made with all his heart and all his soul, will be publickly acknowledged, he wishes for the reasons he could name, that it were to be in Saunders, the Freeman, and the Correspondent and to be from *MR* Thomas Pleasants of William Street, not to ESQUIRE him, for without affectation, he long since took such aversion to the word that none of his acquaintances used it to him. He owns it to an oddity. It is requested that it may be distinctly mentioned, a gift of six thousand pounds, 4 (thousand) of them for the grand operating room and its appendage ones. The interest of the 5th (thousand) annually or as they may be wanted, old linen, lint etc. for such as may be operated on in that room, and the interest of the 6th (thousand) to be expended on wines, spices, and whatever in the cordial way may be wanting for the patients in that room.

NB The money is paid by his representative Mr Jackson Pasley, merchant, £6,000 in Bank of Ireland notes.

Having had estimates of the cost of extending the hospital in the Coombe, and having received this substantial donation from Mr Pleasants, the standing committee decided that it would be better to build a new hospital on a site with space around it than to extend the Coombe building. It was decided to acquire the three-acre site at Long Lane which was owned by the Dean and Chapter of St Patrick's Cathedral. Jonathan Swift had originally leased this area in 1725 for his own private use, 'to enclose a field for horses, being tired of the knavery of grooms, who foundered all my horses and hindered me from the only remedy against increasing ill-health', as he wrote to Alexander Pope. He had built a strong wall around it, about eight or nine feet high, upon which he spent over £600, 'squandering all I had saved on a cursed wall'. Some of the wall is still believed to exist, and a few fruit trees survived at the west end until the 1960s. Swift had called it Naboth's Vineyard. However, the biblical name was dropped and it was generally known as the Dean's Vineyard.[15] In April 1815 the governors wrote to Thomas Pleasants, who had given the money to enlarge the hospital building in the Coombe, to explain that a decision had been made instead to build a new hospital in an open site, and to seek his approval. His somewhat verbose and quaint reply, through his representative, went as follows:

> Mr Pleasants is not well able to write having a sprained thumb of long standing, and the more he advances in life the more he feels it; however the pain is, at present beguiled by the pleasure of returning thanks to Mr Latouche, Mr Maquay, and Mr Guinness for the politeness of their annexed letter, not forgetting the rest of the gentlemen of the Committee.
>
> When he sent the money, and he did it with all his heart, he was confident of its being expended to the fullest advantage. There was in the Committee skill and will for it, and they are, out of question taking the best, the very best step by removing from a confined, low, dirty, bad-aired situation, to no doubt, an open elevated and good-aired one,—of all buildings hospitals demand pure air. He looks on it, that if an hospital had accommodating apartments of every description, the benefit of the most able professional abilities, and the laborious and unremitting attentions of the Governors, yet after all wanting wholesome air, it might be said to be only going on upon crutches.
>
> To give more than verbal proof of his approbation of them getting, and as quickly as they can, from where they are, he will contribute an hundred pounds toward the expence of moving.
>
> What a picture of misery, and what a blessing to prevent it. Some author says (we'll all agree that it is not Bonaparte, author only of infernality!), 'Let one like to see a fine house, another a fine coach, another a fine horse, let me see an happy face.'

> Mr Pleasants presents his good wishes to Mr Guinness. He was intimate with his estimable father: thinks Dublin could not boast of a more valuable character; and in manner and deportment, always the gentleman. And from what he hears, his sons inherit his virtues.

His reference to a 'Mr Guinness' was almost certainly the 'second Arthur', Arthur Guinness (1768–1855). His father, the 'first Arthur', died in 1803.[16] The gibe at Napoleon is interesting as there had been fear of invasion in Ireland in those times and the defensive Martello towers had been completed only a few years before. However, the battle of Waterloo in June 1815 ended that worry.

Thomas Pleasants was born in Carlow and inherited property at the north end of Capel Street. In 1787, when he was nearly sixty, he married Mildred Daunt, the daughter of George Daunt, a well-known surgeon of Mercer's Hospital. They had no children. They lived at 23 South William Street, his father-in-law's house. Mildred died in February 1814 and left a substantial fortune of £90,000. Thomas, who lived until 1818, was always charitable. His first enterprise was building a Stove Tenter House in Cork Street, in 1814, for the use of the woollen weavers to dry their fabrics, at a cost of £10,000. In January 1815 he was made an honorary member of the Dublin Society (now the Royal Dublin Society) on account of his extraordinary munificence to the manufacturers of this city. He also made a donation to the Botanic Gardens, which was used to make the entrance and the porter's lodge. His will was 32 pages long and it included bequests to Cork Street Fever Hospital, a school, an orphanage for girls in Camden Street and of oil paintings to the Royal Dublin Society.[17] There is a large portrait of him on the first landing of the Members' Rooms at the Royal Dublin Society in which he is seated, holding a plan of the entrance to the Botanic Gardens. One of the streets near the present hospital between Heytesbury Street and Camden Street is named after him. It is fair to say that, if it were not for Thomas Pleasants, the hospital would not have been built at Long Lane. He did not live to see the hospital opened.

1815 Act of Parliament
In July 1815 a copy of the new Act of Parliament under which the Meath Hospital and County Dublin Infirmary is constituted was laid before the standing committee (*see* Appendix I). The immediate importance of the Act at that time was to empower the governors to purchase land and spend money in relation to the new hospital. It also contained important sections on the governors, election of the standing committee, and the election of

medical staff, which are highly significant in the subsequent history of the hospital.

The governors, now armed with the approval of Mr Pleasants for the revised use of his money and having raised about £1,600 from the general appeal, made repeated applications to the Grand Jury of the county for a grant towards the new building. An appeal was also made to the Chief Secretary Robert Peel, in January 1819, to use his influence with the Grand Jury. A grant of £5,000, to be paid in instalments, was finally agreed by the Grand Jury, enabling definite building plans to proceed.

Meanwhile, in November 1816, Councillor Huband advised the standing committee to petition the Lord Lieutenant to open Long Lane, which ran along the front of the new hospital site. Accordingly, a petition describing Long Lane as 'a passage leading from Kevin's Park to New Street' was sent, which stated:

> that a large and commodious building is shortly to be erected for a New County Infirmary on the south side of Long Lane, that the said Long Lane is at present impassable owing to vast accommodation of rubbish which has been deposited on its surface during a series of years past whereby the level has been raised many feet above the adjoining ground and not part of it is either gravelled or paved. A plea is made to widen and gravel or pave the lane.

There were many delays in the building of the new hospital before the transfer could be made from the Coombe to Long Lane. In February 1818, it was reported to the governors that some 'depredations' had been committed at the new site and a quantity of lead had been carried away. The angry committee examined Patrick Fox, the watchman, who stated that 'not having arms it was impossible for him to protect the concerns. He could not undertake the care and protection of the place any longer as he would be afraid of his life.' The committee resolved 'to purchase a good musket and bayonet and a pair of pistols to arm the watchman'.

Whitley Stokes

The earliest minute book of the Meath Hospital, found during the preparation of this history, contains entries from the medical board dating from 1776 to 1822 and covers the period up to the time the hospital transferred from the Coombe to Long Lane. On 14 December 1818, at a meeting of the medical board—comprising Dr Harkan and surgeons Dease, Hewson and T Roney, with Solomon Richards in the chair—Dr Whitley Stokes was elected physician to the hospital, to fill the position of the late Dr Egan.

Whitley Stokes had gained fellowship of Trinity College at the age of 25. He was a member of the United Irishmen but resigned in 1792, when the organisation moved from a constitutional to a revolutionary role. He remained nationalist in outlook, however, and, because of his sympathies, his fellowship was suspended by Trinity College for one year. He studied medicine in Dublin and Edinburgh and graduated in 1793; he was elected an honorary fellow of the College of Physicians in 1816. His arrival at the Meath Hospital was the beginning of a long association of distinguished members of the Stokes family with the institution. Apart from his scholarly distinction, he was known for his kindly and charitable nature. He remained physician to the Meath for eight years and resigned, at the age of 53, in favour of his son William Stokes. As will be seen, after his resignation he continued to work in the huts in the grounds of the hospital during the fever epidemic. He had been previously professor of medicine in both Trinity College and the College of Surgeons, and was elected regius professor of physic in 1830.

Robert Graves
In 1821, eighteen months before the move to the new building at Long Lane, Robert Graves was elected physician, in place of Dr Harkan, who had resigned. Patrick Harkan from Roscommon had originally studied for the priesthood for several years in Rome, but he had changed to medicine and graduated in Edinburgh. When he returned to Dublin he was appointed to the Meath as physician, but he remained there for only four years. His letter of resignation, recorded in the minutes of 21 June 1821, did not reveal any discord, stating, 'In separating myself from my colleagues collectively, I beg leave to say that I shall ever feel a warm interest in the prosperity of the hospital, the professional duty of which they perform with so much zeal and ability.' He was also appointed to the Cork Street Fever Hospital and worked there during major typhus epidemics for up to forty years.[18]

Graves' appointment was probably the most important one in the history of the hospital, as he later brought international fame to the hospital and to Irish medicine. However, on 21 July 1821, before the election, the standing committee sent the following communication to the medical board.

> The Standing Committee having heard a report that a bargain has been made between two Medical Gentlemen whereby a consideration in money was to be paid on the appointing of physician to fill the present

vacancy and as the Committee are of the opinion that any proceeding of such a nature will be injurious to the welfare of the hospital and that should the measure be pursued it will become their duty to resist it by all means in their power. The Committee are therefore desirous to call the attention of the Medical Gentlemen to the subject convinced that they will feel the expediency and propriety of preventing any election under such circumstances.

The medical board replied that 'they did not feel competent to institute an enquiry into a subject that rests on no better foundation than a report or conjecture' and that 'they are determined to be guided in their choice of a medical officer solely by a consideration of his moral and professional qualifications'.

The minutes of the medical board for 31 July 1821 resolved that the following communication be sent to the standing committee:

The Medical Board beg leave to announce to the Standing Committee that they have elected Doctor Robert Graves in the room of Doctor Harkan, and the Medical Board take the opportunity of assuring the Standing Committee that in electing a Gentleman of Doctor Graves' character and qualifications they conceive they have considered the best interests of the hospital. Referring to the communication of the 30th inst the Medical Board beg leave most distinctly to declare that the charge relative to 'the purchase and sale of the office' if by that is meant any transaction of a pecuniary nature between the candidate and the electors is without the slightest foundation.[19]

Chapter Two

The New Hospital, Graves and Stokes, 1822–5

Transfer of the patients from the Coombe to the new hospital at Long Lane took place on Christmas Eve 1822. The patients were carried in long baskets made for the purpose, supervised by the surgeons William Henry Porter and Maurice Collis. There was a severe storm at the time and, on the return journey, they protected their heads from flying slates with the empty baskets.[20] The new hospital opened on Christmas Day and, as well as Porter and Collis, the following were the medical staff: Philip Crampton, Cusack Roney, Thomas Hewson, Thomas Roney, Whitley Stokes, Rawdon Macnamara and Robert J Graves.

There is no record of an official opening ceremony at the new hospital. Widdess attributes this to the fact that the newspapers at that time were filled with protests against 'the Orange outrage'. This had occurred during a command performance of *She Stoops to Conquer* at Hawkins Street Theatre, which was attended by Marquis Wellesley, the Lord Lieutenant. The performance had been marred by interruptions from the 'one shilling gallery'. At one point a quart bottle was thrown at the conductor, who stopped the orchestra and held up the bottle. Finally, a watchman's rattle, a substantial piece of timber, hit the cushion in front of his excellency and rebounded between his head and the chandelier of his box. This was the work of Orangemen, who were disappointed by the Marquis' decision to forbid the annual custom of dressing the statue of King William III because it was likely to cause a breach of the peace.

Graves and Stokes

William Stokes succeeded his father, Whitley, in 1826. This resulted in having on the staff the famous combination of the two physicians, Robert Graves and William Stokes. On his return from his travels on the Continent, Graves introduced a system of bedside teaching seen in Germany, in which the students were to study and record the progress of patients assigned to their care under the supervision of the physician. Arthur Guinness, a former student of the Meath, described Graves:

> He was a remarkably fine, tall man, dark complexion and hair. He was in my time the most hard-working of all the staff. As he had a very large practice he used to come in winter time, when I was resident, about 7 o'clock in the mornings when it was quite dark to visit the wards, and many a time have I walked round with the Clinical Clerk Hudson, and often carried a candle for Dr Graves.[21]

William Gibson, Professor of Surgery in the University of Pennsylvania and surgeon to the Philadelphia Hospital, visited Dublin in 1839. He wrote:

> Among the most eminent of the Dublin physicians I may enumerate Dr Graves—so advantageously known among us by his writings on clinical medicine. It was with peculiar satisfaction I found this gentleman still at his post, especially as Drs Colles, Macartney and Stokes had taken advantage of the August holidays and run over to the Continent—a trip made once a year by many of the most eminent British physicians and surgeons; instead of going, as formerly, for recreation, upon shooting or fishing excursions, for five or six weeks, to the moores [*sic*] and mountains of their own island. I had heard nothing it so happened, of the appearance or manners of Dr Graves, and therefore, when ushered into his presence in his own parlour, was startled at finding myself looking up to the face of a very tall, slender, handsome, and well-dressed man, whom, if I had met by chance in the street, I should have turned round to look at and inquired of some passer-by who he was. First impressions are said to be generally the best, and I at once settled in my own mind, that Graves and I would soon be on intimate terms; for I quickly perceived a certain something in his keen, penetrating eye, arrangement

of the muscles of his face, and mobility of the tip of his long and delicately-formed acquiline nose, plainly indicative of humour, high spirits, and quizzical propensities, with ample power to subdue or bring them forward at pleasure.

After this I saw as much of Dr Graves as a fortnight in Dublin would allow. That is, I saw him every day, either at his own house, or at my lodgings, or at the Meath Hospital, where he may be found, every morning, peeping and prying into every hole and corner of the building cracking jokes with the patients or pupils, or old women, or poring over, intently, some medical production, or volume of natural history, or book of travels; for he is very fond of such studies, and took great delight in asking all sorts of questions about our Indians, and lakes, and trees and prairies, and cataracts and great rivers, and buffaloes. And, from all I did see of him, I felt justified, I thought, in jumping to the conclusion, that he is a man of very extraordinary abilities, sufficient to enable him to master any subject to which he may devote his attention, and to sift, thoroughly, any medical case and follow out, successfully, its treatment; but that he is seldom capable, for any length of time, of devoting himself, soul and body, to any given point, or subject; that he is too fond of analogy and of drawing conclusions from solitary facts, so that his listener is always left in doubt as to the certainty of his deductions, however striking and brilliant they may be, as they generally are, from wanting full confidence in his premises.[22]

William Stokes wrote the first book on the stethoscope in English, *An Introduction to the Use of the Stethoscope*, while still a student at Edinburgh; it was published in 1825, the year he graduated. A year after returning to Dublin he was appointed, at the age of 22, to the Meath, in place of his father, Whitley.

Both Graves and Stokes were excellent teachers and, through their publications, became well known internationally. Graves was to retire in 1843 at the unusually early age of 47, for obscure reasons. So the golden period when both were colleagues at the Meath lasted for 17 years, from 1826 to 1843. During this period, in 1835, Graves published his classic description of thyrotoxicosis [23] and his *System of Clinical Medicine* in 1843. Stokes published his *Diseases of the Chest* in 1837 [24] and his description of slow pulse and syncope in 1846, now known as Stokes–Adams Syndrome. His most famous book, on diseases of the heart and aorta, was published later, in 1854.[25]

The teaching and writings of Graves and Stokes had great influence throughout the medical world. There was a marked increase in the number of students attending the Meath in the 1830s and many medical graduates visited from abroad, including North America. Coakley has drawn attention to a number of students who came under the influence of Graves and Stokes who reached great distinction in later life. These include: William Wilde, pioneer eye and ear surgeon; Arthur Leared, inventor of the bi-aural stethoscope; Robert Bentley Todd, neurologist and founder of King's College Hospital, London; Thomas Andrews, medical scientist and vice-president of Queen's College, Belfast; Robert Kane, founder, with Graves, of the Dublin Journal of Medical Science in 1832 and, later, president of Queen's College, Cork.[26]

The great physician, William Osler, acknowledged the influence Graves and Stokes had had on him, through his teachers James Bovell of Toronto and Palmer Howard of McGill University, Montreal. At the bicentenary meeting of the School of Physic in Dublin in 1912, he said, 'I owe my start in the profession to James Bovell, kinsman and devoted pupil of Graves, while my teacher in Montreal, Palmer Howard, lived, moved and had his being in his old masters, Graves and Stokes.'[27]

Professor Armand Trousseau, the famous clinical professor in the Faculty of Paris, also paid tribute:

> I have constantly read and re-read the work of Graves; I have become inspired with it in my teaching.
>
> Shall I now say what are, in Graves' work, the most remarkable and important lectures? To be just, I ought to indicate all in succession: there is not one of them, in fact, which does not abound in practical deductions; there is not one which does not bear the impress of the admirable and powerful faculty of observation which distinguishes the physician of the Meath Hospital.
>
> There is not a day that I do not in my practice employ some of the modes of treatment which Graves excels in describing with the minuteness of the true practitioner, and not a day that I do not, from the bottom of my heart, thank the Dublin physician for the information he has given me.
>
> Graves is, in my acceptation of the term, a perfect clinical teacher. An attentive observer, a profound philosopher, an ingenious artist, and an able therapeutist; he commends to our admiration the art whose domain he enlarges, and the practice of which he renders more useful and more fertile.[28]

Other Medical Staff

While there have been several biographical studies of Graves [29] and Stokes,[30] less attention has been given to the surgeons who worked at the Meath during the same period. There were six surgical appointments, compared with two for physicians.

Philip Crampton was appointed surgeon in 1798, in succession to William Dease, and was on the staff of the Meath for sixty years. Widdess summarises: 'four times president of the College of Surgeons, accomplished comparative anatomist, founder of the Dublin Zoo, skilful diagnostician and operator, he introduced the Meath tradition of genito-urinary surgery by his practice of lithotrity and lithotomy'.[31] He received many honours; he became surgeon-general in 1813 and was created a baronet in 1839.

Gibson of Philadelphia, on visiting Dublin, wrote:

> From all quarters I had heard of his high reputation and enviable fame, and was not, therefore, surprised upon reaching Dublin at the constant exclamation: You have seen, of course, our Napoleon of surgery.
>
> Independently, however, of native intellectual and acquired professional advantages, there can be no doubt that Sir Philip's great success and reputation may be attributed, in some measure, to personal attraction, and to refined and elegant manners; for of all the individuals it has been my lot to meet in European or American Society, no one, in such endowments and accomplishments, has approached so near the standard of excellence or perfection.[32]

Crampton was of striking appearance, 'an active well-built man'. He dressed flamboyantly, sometimes coming to the hospital in hunting dress; in spring and summer he would wear a blue tail-coat with gilt buttons, a white waistcoat and white trousers, and top boots. His nickname was 'Flourishing Phil'. He died on 10 June 1858 and, according to his wish, his body was encased in Roman cement, in the presence of surgeons Rynd, Josiah Smyly and P C Smyly, and was interred in Mount Jerome cemetery.

As might be expected from his extrovert nature, Crampton was a popular teacher. William Henry Porter was the other surgeon at the time, who also had a reputation as a fine, eloquent teacher. He was professor of surgery at the College of Surgeons and president in 1838. The presence of surgical teachers such as these, coinciding with the time of Graves and Stokes, assisted in bringing together the teaching of medicine and surgery at the same clinical school. Prior to this, surgery had been looked upon as an inferior subject. From the early years of the Meath Hospital, the main system of instruction in surgery was by apprenticeship.[33] The term of apprenticeship was five years, at the end of which a certificate was issued. It was generally acknowledged that those who had done their apprenticeship at the Meath would be in a favourable position to be appointed later as surgeon to the hospital.

Thomas Hewson, surgeon for twenty-two years from 1809, had been apprenticed to Solomon Richards in 1800. He was regarded as 'a skilful surgeon and an agreeable companion'. He was elected president of the College of Surgeons in 1819. The brothers Cusack and Thomas Roney were both surgeons on the staff. They were the sons of Patrick Cusack Roney, who also had been surgeon to the Meath and to whom both sons had been apprenticed. Cusack served the hospital for forty-seven years and was president of the College of Surgeons in 1814 and 1824. His younger brother, Thomas, died after twelve years on the staff and was succeeded by Maurice Collis in 1825. Rawdon Macnamara (primus) from County Clare had been indentured to Sir Philip Crampton and had been demonstrator of anatomy with Professor Macartney in Trinity College. He was appointed in 1819 and was president of the College of Surgeons in 1831. He published works on foreign bodies in the trachea. He died of fever in 1836.

Financial Concerns
While the Meath Hospital at the time of Graves and Stokes had such outstanding surgeons on the staff, most of whom became elected presidents of the College of Surgeons, the hospital was still having financial problems. The plain stone building, which remains the central building of the present-day complex, is rather impractical. The steps to the first floor lead to the entrance hall, and for many years this floor was used for accidents and administrative offices. The second floor was surgical and the top floor was medical and was reserved for fever cases for many years. The kitchens were on the ground floor.

Erinensis gave a contemporary description [34] of the new hospital in one of his satirical commentaries in the *Lancet*:

> ... if summer you arrive at a neat plot of ground, smiling with green like an oasis in the desert, surrounded by a high wall, having two entrances guarded by gates of massive iron. Start not reader, we are not about to introduce you to an enchanted castle,—For in the centre of this enclosure stands, fresh from the trowel, the New Meath Hospital and County Dublin Infirmary.
>
> ... a pompous pile of steps, by which you ascend and find yourself landed at once in an apartment which serves the common purpose of a

hall and a waiting room for entering patients, a portion of it being railed off for a surgery where ulcers are dressed and operations performed.

The building was ultimately intended for 100 patients, but only 60 beds were made available when the hospital opened. Later, the standing committee reduced the number of beds to be used to 50, as there were not the funds to maintain more—a situation which seems to have echoes to this day. A cynical letter from the medical board to the standing committee read as follows:

> Assuming as an acknowledged fact that want of accommodation existed in the old hospital, it seems curious that so very slight an addition should have been made to the number of patients on the opening of the new, or that there should have been a necessity for expending several thousand pounds in the erection of an edifice which conferred little more advantage on the poor of the country than the original one, so much complained of. It stands, therefore, as a practical anomaly that the metropolitan county of Ireland should have built a hospital three-fourths of which are unoccupied by patients, for either such an edifice was useless and should not have been erected, or it is useful and ought to be supported to its utmost extent. The interests of the charity should also come under consideration, for not a single day passes that they are not obliged to refuse admittance to numbers of miserable wretches who are then thrown back upon the world to struggle both with poverty and the visitation of disease.[35]

The financial management of the hospital was not easy. In 1825, to supply the range in the kitchen and the fireplaces for the winter, 70 tons of 'the best Whitehaven coal' was ordered by the standing committee. The minutes refer to the problem of where to store such a large amount of coal. Efforts were made to be economical, and a table was issued for the distribution of coals for each room in the hospital. For example, for the months of January, February, March, October, November and December, the resident pupil was allowed one ton and the apothecary a half ton. For the months of April, May and September, the resident pupil got a half ton and the apothecary three-quarters. No coal was issued for June, July and August. Similarly, in January, February, November and December, the resident pupil received seven candles and the apothecary ten. For April, May and September, the pupil received five and the apothecary seven. For May, June, July and August, the pupil received three and the apothecary four. The hospital did, of course, also generate some income; we are told that in June 1825, the grass that had been cut around the hospital was sold for £2!

The monthly salaries and wages of the hospital staff were listed in the minutes of the joint committee as follows:

Ed Mathews, Register	£20 0s 0d
H P Bell, Apothecary	£12 10s 0d
Mary Maiben, Housekeeper	£12 10s 0d
Thomas Fielding, Porter	£5 17s 0d
Lowry Graham, Porter	£5 17s 0d
Susan Ormsby, Nurse	£2 15s 0d
Eliza Gorman, Nurse	£2 15s 0d
Eliza Walsh, Nurse	£2 15s 0d
Mary Gardiner, Nurse	£2 15s 0d
Mary Donohoe, Cook	£2 3s 7d
Mary Tobin, Laundress	£2 9s 4d
Cath Howard, Housemaid	£2 3s 7d

In October 1825 the housekeeper complained that the work was too heavy for one maid and four nurses; the apothecary, Mr Bell, was asked to be released as he wished to go to Edinburgh to study medicine; and in November Mary Tobin was committed to Newgate prison for pledging house linen. The housekeeper was also reprimanded for not supervising the laundry efficiently. Such administrative problems were dealt with by meetings every two weeks of the standing committee and the register, who from the 1840s was called the 'registrar and purveyor' and, more recently, 'secretary'.

An advertisement was issued for an apothecary, stipulating that he must have passed the Apothecaries Hall, be unmarried, and must live in the hospital. Nine applications were received and D Pakenham was elected by ballot. On 21 November 1825, the costs of 'medicines' for the hospital were as follows:

Bernard Fegan, Oatmeal	8s 6d
Boileau and George, Drugs	£7 18s 3d
Fowler, Castor Oil	£2 12s 0d
W Fitzgerald, Lint	£1 1s 0d
Jos. Townsend, Leeches	£1 5s 9d
Denis O' Brien, Lard	11s 8d
W McDonnell, Paper	15s 8d
W Connolly, Wine	19s 0d
Eades, Spirits	£1 7s 0d
Birch, Tow	10s 2d
Ed Allen, Linen and Calico	£2 12s 0d

While all this stringency and day-to-day management was taking place, there was an unexpected and welcome windfall in the shape of a legacy of £2,000 to the hospital from the late Dr Barrett, vice-provost of Trinity College. 'Jacky' Barrett was a famous eccentric in Trinity, a classical scholar who was regarded as being not of this world: '. . . he was so ignorant of the things of common life, that he literally did not know a live duck from a partridge; and though he had dined at commons on mutton for forty years did not know a live sheep when he saw one'.[36] He was said to be extremely miserly, but by his bequest he showed that he had a heart and was aware of the needs of his less fortunate fellow men.

CHAPTER THREE

Fevers and Famine, 1826–50

On 9 June 1826, a letter from William Gregory at Dublin Castle was received by the standing committee.

> It having been represented to the Lord Lieutenant that additional accommodation for fever patients can be afforded in the Meath Hospital to the extent of 24 beds, I am directed by his Excellency to request that you will cause such accommodation to be afforded, and direction will be given for defraying the expense that may be thereby incurred.[37]

The committee were not slow to note that special funds would be available to the hospital for supplying these extra beds. A prompt reply was sent to Mr Gregory the following week, on 13 June, stating that 'the entire upper floor of 36 beds can be made available'. It soon became apparent that more facilities would be needed for the huge number of fever cases. Cork Street Fever Hospital and the Hardwicke Hospital were overflowing.

The majority of cases of fever were typhus. The name typhus comes from the Greek word for 'cloud' and refers to the clouding of consciousness and delirium which is a common feature of the disease. Some of the cases with abdominal symptoms and a longer duration of fever were distinguished clinically as 'typhoid' as early as 1847. The cause and the mode of transmission of these diseases were not known, and malnutrition, atmospheric changes and poor drainage were among the factors discussed by Graves.[38] It was not known that typhus was transmitted by the body louse until 1910, when Howard Ricketts in the USA made the discovery.[39]

The typhoid bacillus and its spread by contamination of water or food was discovered in the 1880s. The spread of typhus in crowded conditions when lice are common can now be understood. Although Graves did not know exactly how typhus was spread, he recognised that crowded conditions favoured the spread of the disease.

> ... but never perhaps in the history of the world was such a fearful commentary on the effects of the entassement of individuals witnessed as in Ireland during the year 1847.
> ... that the Irish epidemic of 1847 had its origins in that congregating together large masses of people at public works and at depots for the distribution of food, and in the overcrowding of the workhouses.[40]

On 8 July 1826, the following letter from the Board of Health was sent to the Meath.

> I am directed by the General Board of Health to request you will inform them whether you will permit sheds of temporary framework to be fitted up in the temporary accommodation of fever patients to the extent of two hundred.
> The building to be erected and the patients supported at the expense of the Government.

The standing committee agreed and extra staff and supplies were ordered at the expense of the Board of Health. There would now be more patients in huts in the grounds than in the hospital itself.

Two extra porters, equipped with white overalls, were to be employed. Part of their duties was to operate a special horse and 'carriage with springs' for conveying patients to the fever sheds. Four laundry maids were employed, for steeping and washing clothes, and an assistant cook. Apart from the four large fever sheds providing accommodation for 50 patients each, sheds were also constructed to house the laundry maids and the horse and carriage.

A list of special supplies included:
60 tins for labels to hang at the patients' beds
1 cwt soap
8 doz rush lights
1.5 stone potash

2 pewter syringes
8 pewter measures
3 straight waistcoats
1 large tub for strong oatmeal
2 flummery tubs
1 canvas bier for carrying the dead
miscellaneous chests, boxes, baskets, cutlery, etc.

Flummery is defined as 'food made by boiling oatmeal down to a jelly'. Note that the list includes 'straight waistcoats' for restraining patients in delirium, which is common in typhus. Graves refers to this measure in describing management of a case of severe fever.

> It will be sufficient to observe, that when he came under our care the chief features of his case were delirium, accompanied by total want of sleep, and a violence of conduct and behaviour calling for the restraint of the strait waistcoat.[41]

Special salaries were paid by the Board of Health for extra duties. A list in the minutes of 4 December 1826, shows that the doctors were paid 5 shillings a day.

Dr Robert J Graves	1 July–4 October	£24 0s 0d
Dr William Stokes	ditto	£24 0s 0d
Dr Whitley Stokes	8 August–4 October	£14 10s 0d
Dr G A Kennedy	ditto	£14 10s 0d
Edw Mathews, Register	1 July–4 October	£15 12s 0d
Danl Pakenham, Apothecary	ditto	£15 12s 0d
Mary Maiben, Housekeeper	ditto	£10 8s 0d

The Board of Health's pay of five shillings a day to doctors attending patients in the fever epidemic was considered quite inadequate for difficult and dangerous work. Over a thousand practitioners signed a memorial to the Lord Lieutenant, pointing out the fearful mortality from fever among the medical men of this country, 'we have to deplore the loss of many of the best and most efficient practitioners who contracted typhus fever in the discharge of their duties among the sick poor. We must strongly and respectfully protest against the five shillings per day by the Board of Health for the discharge of that onerous and dangerous duty.'[42]

The conditions in the fever huts suffered seriously from overcrowding.

Constantia Maxwell quotes one of the doctors at the Meath Hospital: 'Sheds were built, canvas tents were erected, their floors covered with hay, on which the crowds of patients conveyed to the hospitals in carts were literally spilled out. I have seen as many as ten patients lying on the hay waiting their turn to be attended to.'[43]

In October 1826, Graves wrote to the standing committee to direct attention to the state of the floor in number 4 shed: 'It is worn away particularly between the beds where the soft earth absorbing the filth proves a source of most offensive and dangerous exhalations. It is proposed to renew the floor by pieces keeping six beds empty until it is finished.' The committee ordered that 'the Register do forthwith employ a labourer to prepare mortar with hot lime and best fresh water, sand and coal ashes and (if it to be had) forge dust and after removing wet earth and replacing it with dry earth mixed with coal ashes to lay a new floor in pieces where required.'

Graves had had previous experience attending fever cases, in the West of Ireland in 1822, when there was a high mortality among the attending doctors. Graves attacked the Board of Health for its management and low pay to doctors. This included a severe criticism of the distinguished Dominic Corrigan, who was a prominent member of the Board.[44] Graves was annoyed that the government had not asked for nominations to the Board from the College of Physicians or College of Surgeons. Corrigan was not a fellow of either college and later he was black-balled at an election to the honorary fellowship of the College of Physicians. However, he was to be elected president of the College in 1859, the first Catholic to achieve that office, and was made a baronet in 1866.

In August 1826, the Board of Health wrote to the Meath stating that, during the fever epidemics of 1817–19, 'advantage had been experienced' by distributing soup and bread to convalescent patients after they had been discharged from hospital. They requested that the Meath put 'similar measures into effect on the present occasion'. The Meath replied that in 'the present crowded state of the grounds of the hospital, the vast number of fever patients, in addition to the ordinary patients and attendants, the governors could not with safety or convenience undertake the distribution of soup and bread on the premises'.

However, the Meath had issued tickets for meals to be obtained by convalescent patients at the Dorset Nourishment Institution and they quoted a reply from Anne Bewley when this had been arranged: 'I have been at the Soup Kitchen and proposed to our people there supplying the patients dismissed from the County Infirmary with broth and bread. Our boilers are very large and I think adequate to supply 200 persons daily.'[45]

Diet of Fever Patients

The accepted treatment of fever patients was to severely limit the diet. At the beginning of the epidemic, in July 1826, the joint committee had resolved that:

> the following dietary recommended to the sub-committee by the medical officers be adopted:
> 1. Low: Two quarts of whey.
>
> 2. Convalescent: One quart of whey, 4 oz bread & 1 pint milk—morning,
>   Flummery in the evening.
>
> 3. Small Middle: 4 oz bread & 1 pint milk—morning,
>   4 oz bread & 1 pint milk for dinner,
>   Flummery in the evening.
>
> 4. Middle Bread: 8 oz bread & 1 pint milk—morning,
>   8 oz beef and broth for dinner,
>   Flummery in the evening.
>
> 5. Full diet: 8 oz bread & 1 pint milk for breakfast,
>   8 oz beef, 8 oz bread (or 2 lb potatoes),
>   A slice of ox cheek & pint of broth for dinner,
>   Flummery in the evening.

However, Graves believed in feeding fever patients adequately and not keeping them on a scanty diet, so it seems unlikely that his patients were restricted to the low diets described above. By better nourishment of his fever patients he achieved much improved results. He suggested to his students that if they wanted an epitaph for him it could be that 'he fed fevers'.

On 13 November 1826, with the onset of winter, the standing committee resolved that each shed be furnished with two stoves, 30 inches high and 14 inches in diameter, lined with firebrick and furnished with flues 6 feet high— 'by burning coke no smoke is produced and the warm air circulates through the ward'. Sheeting of thin boards was also fastened to the outside of the sheds.

New Fever Department

Towards the end of 1827 the numbers of fever cases declined and the Board of Health advised that the huts be closed. But there was still demand for fever cases to be admitted from Dublin county and it was agreed to continue to reserve the 36 beds on the top floor of the building for fevers.

On 22 November 1828 it was noted in the minutes that Drs Graves and Stokes had not yet received any salary for their two-year attendance at the fever huts, showing that administrative delay is not new! In December 1829, it was decided to sell the huts and put the money towards building a dispensary. An estimate was received from John O'Connor of £366 16s 10d and he would allow £120 for the sheds. The new dispensary was built along the wall of Long Lane; it had several additions over the years and was the out-patient building until the 1960s.

The first record of the use of gas for lighting in the hospital was in October 1843, when the large surgical wards were lit by gas. So, during their time at the Meath, Graves and Stokes would have attended patients by candle-light in the early morning or in the evening.

Graves resigned in October 1843, after 22 years as physician to the hospital. He died ten years later at the age of 58. After his death, the arrangement agreed with the Board of Health that 36 beds should be kept for fever cases continued. Accordingly, the cost of the fever beds was paid by the government. The average annual cost of maintaining the 64 beds in the infirmary department was £1,200 and the fever department averaged £700.

The famine years, beginning in 1845, brought a stream of cases of fever to overcrowded dispensaries and infirmaries in Ireland. Supplies of food to the hospital were not significantly affected, judging from the hospital minute books, but on 28 September 1846 the standing committee noted that a half pound of bread was to be substituted for two pounds of potatoes and that Indian corn meal was to be purchased for stirabout. In Dublin in 1847, although 2,500 beds were provided, 1,000 more than in any previous epidemic, as many as 12,000 cases of fever applied to the Cork Street Hospital in ten months.[46]

On 22 January 1849, a letter from the Lord Lieutenant was received by the standing committee referring to a report of a select committee of the last session of parliament which recommended a progressive diminution in the votes for the Irish charities, with a view to their final cessation. Following this, it was indicated that withdrawal of the grant for the maintenance of the fever beds was being considered. The standing committee were alarmed at this ominous communication and appointed a sub-committee on 29 January 1849 to prepare a reply to the Lord Lieutenant.[47] The sub-committee was composed of Dr William Stokes, surgeon C Roney and a lay governor, Mr Callwell. From the contents of this reply it is certain that Stokes was the main contributor, as it deals in some detail with the teaching of clinical medicine.

The reply opened by stating that the majority of the patients admitted to the fever wards were from the county of Dublin and were always admitted promptly: 'Thus fever is prevented from spreading among the lower classes in the county districts round Dublin.' It continued:

> The number of beds originally established in the beginning of 1828 was 36 and no change has since taken place in the amount of accommodation. It was arranged that the resident officers of the hospital, namely the Registrar, Apothecary and Matron, should receive remuneration for the increase of duty thus laid on them but that the medical and surgical charge of the patients was to be performed gratuitously. This arrangement has continued up to the present time including a period of 21 years. The increase in the total number of beds in the hospital was of great importance to the interests of the Institution, as it was thus enabled to fulfil the conditions necessary for its recognition as a qualifying hospital under the regulations of the various medical and surgical licensing bodies in the United Kingdom.
>
> The establishment of these wards has proved a most valuable addition to the general medical school of Dublin in as much as the Meath Hospital thus became a school of Dublin in which the surgical as well as the medical students could study fever in its various forms, for although during the pressure of epidemics fever cases have been admitted into several of the great hospitals of Dublin, yet the Meath Hospital has been the only one on this side of the city in which fever wards were permanently open to the surgical student—and for this purpose its utility has been greatly enhanced by its proximity to the College of Surgeons and the University.
>
> The governors have further to state for the information of His Excellency that no pains have been spared to render the instruction afforded in these wards of practical value to the student.
>
> It was here that the method of instruction so long practised with advantage in Germany, was first introduced into this country by Dr Graves,—and its results have been most beneficial. Each patient is given in charge to an advanced student whose duty is to make himself master of all the details of the case, to draw up daily reports, to suggest treatment and perform all the minor duties of a medical attendant while the physician acts as director and instructor.
>
> Premiums from funds subscribed by the medical officers are given to the most diligent students, and a special certificate is furnished to the clinical student at the termination of his course. This document has in many instances proved his most valuable qualification.

By an arrangement adopted for some years past, facility is given to the student to pursue both his surgical and medical studies, so that his attention shall not be distracted, the week is divided into days of medical and surgical instruction, and the clinical teachings in each department are thus kept from clashing.

In this way a description of knowledge is given which it is impossible for the student to obtain by any amount of reading or oral instruction, and he leaves the hospital a practical and experienced man.

The total number of students who have attended the hospital since the commencement of the winter session of 1827 is 1,624 giving an average attendance of 77 per annum. All these men attended the lectures, but if it be admitted that but two-thirds availed themselves of the practical instructions it would leave the number of 1,082 as representing the amount of students sent out from this institution qualified by actual experience to undertake the charge of fever hospitals. In this calculation the complementary admissions to the hospital have not been entered, and it is right to state here that medical officers of the Army and Navy are and have been always admitted to attend the visits and lectures in the hospital free of expense—a circumstance which has proved in many cases of great value to the Regimental Medical Officers on their arrival in Dublin, giving them an opportunity of studying the fever of the country to which the troops under their charge will be exposed.

They have also the gratification of stating that this institution has not only been a hospital of instruction but one of investigation. During the last twenty years the number of original memoirs on medical and surgical subjects which have been published by the officers of the hospital amounts to upwards of ninety, and this is exclusive of large works and minor contributions to scientific societies. Most of these memoirs have been translated into the continental languages and also re-published in America.

The committee of governors feel themselves justified in submitting to His Excellency that the withdrawal of the grant from the fever wards of this hospital would be seriously detrimental to the interest of the medical school of Dublin, and to those of the public at large.

The above communication to the Lord Lieutenant gives an interesting summary of the development and standing of the Meath as a teaching hospital in 1849. It appears to have had great influence, as the financial maintenance for the fever beds was not subsequently reduced.

Ever since the 1826–7 epidemic, when temporary fever huts, providing accommodation for 200 patients, had been erected in the grounds of the

Meath, 36 fever beds had been officially designated and maintained in the hospital building on the top floor, representing one-third of the capacity of the hospital. Although this must have resulted in considerable risk of infection to patients with other conditions, it was not until 1874 that a small separate building with two wards was constructed for infectious cases. Most of the fever cases were typhus, and some were of relapsing fever, both of which were transmitted by lice. The excrement of the lice being rubbed into the skin, or becoming dust which could be inhaled, could transmit infection. The other prevalent disease was dysentery.[48]

Graves discusses contagion in his *Clinical Medicine*, published in 1848:

> The fever wards of the Meath Hospital are by no means crowded, and are both well ventilated and cleanly, while the building itself is placed in the most salubrious part of the vicinity of Dublin, being built upon the site of Dean Swift's garden; and yet it almost invariably happens that when a patient, labouring any other acute, or any chronic disease, is admitted into a fever ward, he gets fever in the course of a fortnight or even sooner. This happens the more surely if the patient is placed in the immediate vicinity of a maculated case. Among the pupils who attend the hospital, the greater number are sooner or later attacked by fever, and the same is true of the porters, laundry-maids, and nurses.[49]

Hospital statistics of admissions, discharges, deaths and costs show that the fever wards, containing 36 beds in the year 1852–3, admitted 436 fever cases, with 29 deaths. The infirmary section, with 66 beds, which included general medical and surgical cases, admitted 803 other cases in the same period, with 46 deaths.

Whereas the physicians were largely engaged in attending fever cases, the surgeons were occupied in dealing with injuries, septic conditions and miscellaneous operative procedures such as amputations. These were done without the benefit of general anaesthesia, which did not reach Dublin until 1847.

More Medical Appointments

After the resignation of Graves in 1843, Cathcart Lees was appointed in his place. He was a diligent and painstaking teacher. He gave lectures on diseases of the stomach and indigestion,[50] so his interests did not clash with those of his senior colleague, William Stokes, the pioneer cardiologist. He resigned in 1861 for reasons of health, after eighteen years of service.

Alfred Hudson, who was physician to the Adelaide Hospital, applied for the vacancy created by the retirement of Cathcart Lees. He had been a student of the Meath and had been a clinical clerk to both Graves and Stokes. Guinness, in his memoirs in Ormsby's history, describes Hudson when he was clinical clerk to Stokes as 'little Hudson, he was the most delicate poor-looking little fellow I ever saw, and dressed very shabbily. He thought only of his work, in which he seemed to throw his heart and soul, and no doubt this was the cause of his great success in after life.' Hudson was elected physician to the Meath in February 1861, and the Adelaide committee congratulated him on 'the most prestigious appointment in Dublin'.[51] He then became one of the two physicians at the Meath, junior colleague to William Stokes who had been his teacher. A small bearded figure, quiet and unassuming in manner, he had an impressive earnestness and thoroughness in all he said and did. Although he remained physician to the Meath for ten years, he remained a faithful supporter of the Adelaide and in his will he left money to found the student prize in that hospital, the prestigious Hudson Scholarship. Ormsby's comment on this was, 'as an old Meath man, this is the only act of Dr Hudson's we cannot easily forgive'. He was elected president of the College of Physicians in 1877 and became regius professor of physic in succession to Stokes in 1878.

Maurice Collis was appointed surgeon in place of Thomas Roney in 1825. He had been apprenticed to Thomas Hewson during his training. A popular figure in the hospital who had many apprentices, Maurice Collis was elected president of the College of Surgeons in 1839. He was very religious and was known as 'Collis the Good'. There is a story that a patient attending the famous Abraham Colles of Steevens' Hospital asked him if he was Collis the Good.

'No,' he replied, 'but I *am* Colles the Great.'

The appointment of Maurice Collis began an association of the Collis family with the Meath Hospital that continued for about one hundred years. His nephew, Maurice Henry Collis, succeeded him as surgeon in 1851. William S Collis, a solicitor and son of Maurice H Collis, was chairman of the governors of the Meath before and after the First World War. His son, Robert Collis, a famous paediatrician and writer, was assistant physician at the Meath in the 1930s.[52] After the death of Maurice Collis in 1852, money was subscribed to erect two wards in his memory. These are the two rooms at the east end of the corridor on the first floor, one of which is now the endoscopy unit. A bust of Maurice Collis is placed over one of the doors. Over the other is an inscription:

> These wards have been added by some of his friends and relatives, in memory of the late Maurice Collis, FRCS, for upwards of 25 years one of the surgeons of the hospital, as testimony of their value for one who showed the fruit of faith in Jesus Christ by a life of devotedness to God and untiring benevolence to his fellow men. 1854.

Two Meath surgeons of very different styles were Francis Rynd and Josiah Smyly. Rynd had been apprenticed to Sir Philip Crampton. He was passionately fond of hunting and Sir Philip had found it difficult to get him to attend to his hospital duties in his early days. He was elected surgeon in 1836, in the place of Rawdon Macnamara I. He had very polished manners, dressed most fashionably and was a great favourite with the ladies. He is best known for first injecting medication subcutaneously. He injected morphia for pain with an instrument without a plunger, the fluid entering the tissues by gravity. This method was the forerunner of the modern syringe for hypodermic treatment. Rynd died suddenly after an altercation following an incident in the street in 1861.[53]

Josiah Smyly had also been apprenticed to Sir Philip Crampton, who was his uncle. He was elected surgeon in 1831. He was a kind and gentle man, conscientious and very religious. His wife was of the same charitable disposition and founded homes and schools for poor children. When she died at the age of 86, she left seven homes and four day-schools, which, in the words of her granddaughter, Miss Vivienne Smyly, were 'free of debt but filled with children'.[54] Josiah Smyly wrote on miscellaneous surgical problems such as fractures of the patella, stricture and strangulated hernia. He died in 1864, when he was president-elect of the College of Surgeons. Two of Josiah Smyly's children were to become distinguished medical men. Philip Crampton Smyly was surgeon to the Meath and William Josiah Smyly was master of the Rotunda Hospital.

Chapter Four

The Smyly Appointment Controversy, 1861

William Henry Porter died in April 1861, leaving a vacant surgical post at the Meath. Philip Crampton Smyly was elected surgeon to the hospital in his place on 13 July 1861. The *Irish Times* ran a strong campaign against his appointment and published a leader on 1 July 1861 which included the following:

> The Dublin hospitals rank with the very highest surgical schools of the United Kingdom. Among the Dublin hospitals the Meath Hospital holds a conspicuous place . . .
>
> One of the surgeonships to the Meath Hospital is vacant and has been so for the last two months. This unusual delay in appointing a gentleman to the post has given rise to much speculation as to the cause. It cannot be that there is any want of candidates, or that there are not among the candidates gentlemen of tried ability and long experience. The prevalent rumour is that the place has been kept open for a son of Surgeon Smyly, a young gentleman who has not yet taken the necessary degrees. We are certain that Surgeon Smyly would not for a moment sanction any arrangement of this kind. He knows, fully as well as any other surgeon, that skill and dexterity are not hereditary; and from the deep and earnest interest he has ever exhibited for the hospital with which he is already so intimately connected, he would not suffer that a novice in practice should hold the important position of surgeon to the Meath Hospital. The

board, too, which contains among its members such men as Stokes, Hudson, Rynd and Wharton etc., would indignantly repudiate a transaction of this description. They could not think of entrusting any department of such an institution to a boy.[55]

The reference to P C Smyly not having the 'necessary degrees' presumably refers to the requirement in Clause VII of the 1815 Act (*see* Appendix I) that all elections of physicians and surgeons to the Meath Hospital must be from members or licentiates of the Royal College of Physicians and Royal College of Surgeons in Ireland. This rule was not always adhered to, as will be seen in the 1910 controversy. Philip C Smyly had passed his MB in Trinity College in 1859 and had studied in Berlin and Vienna in 1860.

A number of letters were written to the paper, following the *Irish Times* leader. Among these was one by 'Senex':

In a leader of your journal of the first of the present month you observe that our 'Dublin hospitals rank with the very highest surgical schools of the United Kingdom'.

Such, Sir, was their position, and some of them, I rejoice to say, still retain their status. With others, however, this is not the case. Unfortunately, two in particular could be named that have for some years back, fallen from their estate. The cause of this has been a system of nepotism which is inherent in them. In neither can any physician or surgeon gain a post, save he be the relative, the intimate, or the umbra of some one of the medical staff; in fact they are family concerns. It is to one of these you alluded in your article of the 1st inst. The 'prevalent rumour' relative to the appointment of a certain individual as surgeon to the Meath Hospital, in the room of the late and lamented Mr Porter, is, I fear, but too true. That the situation has been kept open, up to the present moment for what may be designated a foul job, a short time will make evidence; in fact, that the office has not been filled up to the present, in order that a son of one of the surgeons of the hospital may have time to obtain his licence and become eligible to offer himself a candidate for a vacancy, to which he has already been provisionally appointed, is beyond all doubt the case.[56]

On 8 July 1861, the *Irish Times* commented on the letter of Senex, 'the nom-de-plume of an eminent medical gentleman in Dublin'.

The election is a matter of importance to all who regard the independence of the medical profession and the safety of those for whose benefit the hospital is maintained. The Meath Hospital and County Infirmary, for the institution is both, treats daily 112 intern patients, and relieves about 10,000 extern patients annually. Cases arising from serious accidents are conveyed at once to the Meath Hospital. Its reputation throughout the continent is so great, that every foreign medical student visiting Dublin wishes to be taken first through the Meath Hospital, of which he has heard so much. A government, niggardly in the extreme towards public institutions, grants £700 a year to the hospital in recognition of its merits; and its celebrity as a surgical school is unsurpassed by that of any of our justly famous Dublin schools. We are jealous of the reputation, and watchful of the management of such an institution. It must not be supposed that its offices are hereditary, as it is not imagined that surgical skill can be bequeathed by way of inheritance.[57]

Another letter to the paper, dated 9 July 1861, gave a different point of view. This was signed 'Senior', a nom-de-plume later revealed to conceal the name of surgeon Charles Benson:

The only objection made to the contemplated appointment in this case is, I believe, the youth and inexperience of the favourite candidate; and it is designated a 'foul job', because one of the electors is the father of the aspirant; but if it can be shown that youth and inexperience are not objectionable in this instance, then the respected father may reply as the celebrated Abraham Colles once did on a similar occasion, 'I don't see how his being my son disqualifies him.' Now, I assert, without fear of contradiction, that there is no weight in the objection. Where is the old and experienced man to be got? Is it from another hospital? If so, a Tyro must be appointed to that, or else we must look to private practitioners from whom to select a senior man. But it is a well-known fact that a man may be many a long year in private practice, or, rather looking for it without gaining much experience. A young man, on the contrary, soon learns how to act and how to teach. He is not ashamed to learn from his seniors, and to ask for advice and instruction from his more experienced colleagues. Not to appoint a surgeon to a hospital until he was a good operator and an experienced practitioner, would be to imitate the fond mother, who would not allow her son into the water until he should first learn to swim.[58]

Maurice Henry Collis wrote to the *Irish Times* on 11 July to explain that the delay in the appointment was due to difficulties in arranging a full attendance of the medical board for the election, as certain members were away. The *Irish Times* again came back indignantly, on 15 July:

> The 'family arrangement' respecting the vacant surgeoncy in the Meath Hospital has been completed in contemptuous defiance of public and professional opinion. Mr Smyly junior, who obtained a junior Moderatorship in Trinity College, and a medal for an essay not written under competitive supervision, has been appointed to the post, and the patients and pupils of the Meath Hospital will be under the care of a gentleman who has reached the mature age of 22 years. In all our experience we never heard of so gross a job being perpetrated in such a class of professional society.
>
> By a very peculiar arrangement the Medical Officers of the Meath Hospital obtained the privilege of appointing their colleagues by surrendering for the use of the hospital £100 a year. There are seven electors, and of these four voted for Mr Smyly Junior, and three against him. The four who voted for the young gentleman are Mr Smyly Senior, father of the successful candidate, Surgeon Rynd, Dr Collis and Surgeon Wharton. The three who voted against a nomination which virtually made the appointments in the hospital private property are Dr Stokes, Dr Hudson, and Surgeon Porter. These names will be well remembered in the history of the Meath Hospital.

The defeated candidates were John Morgan, FRCS, surgeon to the Adelaide Hospital; Robert MacDonnell, FRCS, professor of anatomy and physiology at the Carmichael School; Rawdon Macnamara II, professor of materia medica; and Robert Persse White, lecturer in medical jurisprudence. Both Rawdon Macnamara and Persse White were elected surgeons to the Meath at later elections.

Questions in the House of Commons

On 31 July 1861 the *Irish Times*, in yet another leading article, continued the attack:

> They who hoped that the election of Mr Smyly junior, in utter contempt of public opinion, settled the matter, and that for the future silence was to encourage acts of the same kind, will be disappointed. The legislature will be made fully as well as acquainted with the whole proceedings, as the medical profession has been, and it is not unlikely that the scandal

will produce as much indignation of the House of Commons as it has caused among the citizens of Dublin.[59]

It went on to say that Mr Hennessy, MP, intended to call the attention of the House of Commons to the Meath Hospital, 'with reference to the manner in which its officers are elected'.

There seems to have been lobbying of members of parliament by both sides, as shown by the following extract from a letter to Josiah Smyly from Mr Vance, MP:

> The matter [appointment of P C Smyly] came on this evening at the vote for the Dublin Hospitals and Dr Brady made not only an incorrect but untrue statement of the facts, describing Mr Smyly as only 20 years of age and attacking the system on which appointments are conferred.
>
> I rose to reply to him but Mr Cardwell got up at the same time. He was called in preference by the Chairman of the Committee and he refuted the charges brought by Dr Brady to the entire satisfaction of the House. I was not left with a word to say nor could I add anything to Mr Cardwell's defence which was completely successful, especially as he referred to a communication he had received from the late Mr Rynd fully approving of what had been done.[60]

The reference to a letter from the well-known Francis Rynd supporting Smyly's appointment must have had some effect, as he had died tragically only a few days before the Commons debate.

In spite of all the criticism of Philip Crampton Smyly's appointment, he went on to have a very successful and distinguished career. He was elected president of the College of Surgeons in 1878 and was appointed medical attendant to a succession of Viceroys. He had a large practice and was noted for his good manners, resulting in the students giving him the nickname 'Polished Phil'. He was knighted in 1894.[61]

Chapter Five

Hospital Conditions and Staff, 1860–87

In the 1860s, frequent references were made in the minutes of the standing committee to the difficulties of raising enough money for the hospital's running costs. A sub-committee reported that increased expenses were due to the rising cost of meat and wine, and that mutton was more expensive than beef and should be curtailed in future.

Surgeons and Medical Staff
George Hornidge Porter had been elected surgeon in 1849, in place of Cusack Roney, when his father, William Henry Porter, was senior surgeon of the Meath. Porter, who had been indentured to Josiah Smyly, remained in office until his death in 1895, a period of 46 years. He gained many distinctions during his career, including presidency of the College of Surgeons in 1868 and a knighthood in 1883. He was also a Grand Juror of the Wexford Assizes and a member of the Board of Superintendence of Dublin Hospitals.

In contrast, Maurice Henry Collis spent most of his energy within the hospital. Appointed in succession to his uncle, Maurice Collis, in 1852, he was active in establishing the Collis wards already described and in raising money to build the first children's ward in the hospital in memory of Josiah Smyly. The Smyly memorial children's ward was constructed on the floor directly above the Collis wards, at the east end of the second-floor corridor, and was officially opened by the Lord Lieutenant and Mrs Wodehouse on 5 January 1865.

A special Smyly ward memorial committee was set up which continued to raise funds right up to the 1900s to support the children's beds independently from the general hospital funds. In 1972 the Smyly ward was converted into an intensive care ward under the direction of the anaesthetist Dr Peter Morck, with funds from the estate of Chester Beatty. During the conversion two marble bas-reliefs in classical style were salvaged from the walls, one depicting 'morning' and the other 'night'.

Ormsby described Maurice Henry Collis as a bold and dextrous operator. Sadly, whilst excising an upper jaw for malignant disease in the theatre of the Meath Hospital, Collis received a slight wound in his hand from a spicula of diseased bone. This injury developed pyaemic poisoning, of which he died seven days afterwards, on 28 March 1869.[62] His colleagues erected a marble bust of him which is in the entrance hall of the hospital. This is the bearded figure, as opposed to the two other busts in the hall which are of William Stokes and Philip Crampton.

Rawdon Macnamara II, son of Rawdon Macnamara I (surgeon 1819–36), was appointed in 1861. He was professor of materia medica in the College of Surgeons and later elected president. All these posts had been held by his father before him. His students referred to him as 'Big Mac'. He appears to have been a great social success and a good storyteller. 'He can make a dinner party extremely pleasant if he likes.'[63] Shortly after his appointment, he complained to the standing committee about the bad state of the front hall and offered to guarantee money to upgrade it. However, this was not done until 20 years later, when Mathew O'Reilly Dease made a donation to restore the entrance hall in memory of his father and grandfather.

The Dease family are commemorated by the excellent tiling of the entrance hall and the corridor running from the west side through to the boardroom. The engraved glass lantern hanging in the hall is also part of the memorial. William Dease and his son Richard were surgeons to the hospital (1793–8 and 1795–1819 respectively), both having worked in the old Meath Hospital in the Coombe, and William was one of the founders of the College of Surgeons. William's death was very tragic. Apparently, he opened a femoral artery aneurism in mistake for an abscess and the patient bled to death, so he took his own life by opening his own femoral artery. His son, Richard, died of septicaemia contracted whilst performing a pathological demonstration.[64]

James Henry Wharton, surgeon at the Adelaide Hospital, was appointed to the Meath in 1858. He was a good organiser and revised all the medical board records. He was known as 'Conscientious James'![65]

In 1864, William Stokes, son of Dr William Stokes the senior physician, was appointed surgeon, in succession to Josiah Smyly. He wrote widely on surgical subjects and had a reputation as an orator. He was appointed professor of surgery in the College of Surgeons in 1872. In 1868 he resigned from the Meath to take up a post at the Richmond Hospital, but he returned to the Meath in 1888.

In April 1868, James Whitelaw Stronge was elected by the medical board after Stokes' resignation. Stronge had been educated at the Richmond Hospital and Trinity College and had practised for many years in Belfast before returning to Dublin. 'For many years he had been a warm and fast friend of Sir George Porter, and it was mainly by the instrumentality of the latter he was appointed surgeon to the Meath Hospital.' He was 53 at the time of his appointment, but he found the demands of teaching a considerable strain. He died of fever in November 1868.[66]

Stronge's appointment was not in the same pattern as previous surgical appointments at the Meath and after his death another unusual appointment was made, in the person of Robert Persse White. After completing his medical studies, White had spent some time as a gold digger in Australia. He returned to Ireland and was a dispensary doctor in Terenure for many years. He was then appointed to Jervis Street Hospital, but he was persuaded by the Meath to resign this post. Unfortunately, he developed poor health and was treating himself with large doses of morphia before his death in 1879.[67]

In June 1868, Mr Frederick Stokes, a lay member of the standing committee, expressed his concern about the appointments procedure and lodged a notice of motion that 'future appointments should be vested in the General Board'. There is no record of any change following this and so, in December 1868, Frederick Stokes offered his resignation from the committee in a letter to Mr E B Stanley, the registrar:

> Dear Sir,
> I cannot concur in the manner in which the affairs of the hospital are controlled by the medical board, and have too deep a regard for these gentlemen to prolong the controversy. I prefer to retire from a position in which I have endeavoured to the best of my time and ability to promote the interest of the hospital. Please therefore to receive my resignation as a member of the standing committee.
> ... I think that the interests of the hospital, as a charity, are made much too secondary to those of the medical profession, and that the whole power of the hospital rests with the medical board, and none with the committee.
> In the matter of the election of the medical staff, so far from the wishes of the standing committee being consulted, they were not treated with

courtesy, and in the disposition of a post of enormous value, the interests of the hospital in a time of great distress were disregarded without the compensation of men of eminence being appointed.

The expenditure despite of constant representation, has rapidly and continually increased, without important increase of numbers, and with larger income the demands have only been met by repeated overdrafts, sales of stock and borrowing, and when the board took the only effective remedy, closing the wards, the order was disregarded by the medical men.

I dislike the onesided character given to the institution by the total exclusion of Roman Catholics from the medical board, leading me to believe that a gentleman of that persuasion would not be chosen whatever his professional claims . . .

I believe the interests of the poor of the city and county suffer by country and other cases of medical interest being brought in the hospital, and also by the excessive time cases are kept in (often from kindness) at the present time . . .

I think it is objectionable in principle that the medical board in such numbers should also hold authority as members of the SC. The extensive power claimed by the former make it more just that the committee should be wholly non-medical.

. . . I feel grateful for all the kindness I have received, and I intend still to consider myself a supporter of the hospital. I trust that even these, if any there be, who are thankful to get rid of a troublesome critic, will yet give heed to some of the points I have referred to.[68]

The standing committee, despite this broadside, replied that they could not consent to lose his services and that they must decline his resignation. But, although Frederick Stokes did not resign in the end, he was still not content with the appointment system. At a meeting of the standing committee on 5 April 1869, following the death of Maurice H Collis, it was proposed by Frederick Stokes and seconded by Sir J Barrington 'that a respectful communication be addressed to the Medical Board requesting them to suspend the appointment of a surgeon in the room of Dr Collis until this Board have an opportunity of a friendly conference with the Medical Board, for the purpose of endeavouring to place the hospital in a position of independence and increased efficiency'. However, the medical board may not have received this communication, or they ignored it, as, on 26 April 1869, James Wharton, secretary of the medical board, informed the standing committee, 'Gentlemen, I have to announce that Dr Mayne has been elected surgeon to the Meath Hospital in room of the late Mr M H Collis.'

Cholera

Cholera outbreaks occurred in Ireland in the eighteenth and early part of the nineteenth centuries. The records of the Meath do not reflect the 1832 epidemic in Dublin, but the hospital was heavily involved in 1866. Sheds were again erected in the grounds, this time by the Board of Guardians of the South Dublin Union, and some medical students were paid to assist. The hospital statistics for the year 1866–7 recorded approximately 100 deaths due to cholera. Some practical aspects of hygiene appeared in the minutes. It was decided that the only way to deal with patients' clothes was to boil them. The disposal of straw used in bedding also appeared to be a problem, as burning it was causing a nuisance to neighbours of the hospital!

Smallpox, 1871–2

Robert St John Mayne was educated at the Richmond Hospital and the Carmichael School of Medicine. He was the son of Robert Mayne, who had been a distinguished physician at the South Dublin Union and the Adelaide Hospital and who had died in 1864 of typhus, at the age of 53.[69]

On 13 March 1871, Dr R St J Mayne, now secretary of the medical board, wrote to the standing committee advising that 'the Dispensary be got ready for smallpox'. An epidemic of smallpox had begun and the medical board did not want to admit such cases to the fever wards on the upper floor of the hospital. On 31 October 1871, the joint committee consented to erect two temporary sheds for smallpox patients, to be financed by the Guardians of the South Dublin Union. Drs William Stokes and Arthur Wynne Foot, Mr R St J Mayne and Mr R Persse White agreed to attend the patients.

Cases of smallpox continued to be admitted until April 1872, and a special dead house shed was erected. In April, an order was made that no new cases would be admitted except county cases, towards which the Meath as County Infirmary had an obligation. The epidemic of smallpox continued for eighteen months (from April 1871 to September 1872 inclusive). During that period, 88 cases—53 males and 35 females—were treated for the disease, of which nineteen died, a mortality of 21.59 per cent.[70]

Tragically, Robert St J Mayne contracted smallpox from patients he was attending in the dispensary and he died on 16 December 1871, at the age of 28. It seems strange that he had not been protected by vaccination, which was well established at that time. Lambert Ormsby was appointed surgeon in place of Mayne.

The occupation of the dispensary as a temporary smallpox hospital seriously interrupted general out-patient work. It was now so obvious that permanent isolation wards were required that the standing committee decided to build 'epidemic wards' and a laboratory in the grounds on the west side. Building work was completed in 1874, providing two wards, each to accommodate six patients. This was the first provision for infectious cases outside the main hospital, but the extra accommodation was not always sufficient. Cases of smallpox and cholera recurred in 1876, and the Board of Guardians were preparing to erect tents in the grounds.

Review of Hospital Staffing
In August 1873, the joint committee decided to review the staffing situation in the hospital. The registrar—or secretary, as he was now known—was Mr E B Stanley. His salary was £170 per annum and he was allowed 5 per cent of rents collected. He had full accommodation in the hospital, consisting of four rooms and a portion of the grounds. His duties had never been specified in detail, and the committee, though pleased with his work, made certain recommendations. It was to be made clear that he was to have full control over every officer and servant in the house. This included resident pupils, and referred in particular to disorderly conduct.

As regards his portion of ground, it was decided that no area should be diverted from its proper use 'for health and recreation of the inmates'. The ground was not to be used for cultivation, rearing pigs or poultry, or for laundry purposes, and Mr Stanley was obliged to remove these activities from his plot. For this he would be compensated by an extra £5 per annum. Later it was decided that he should be paid £185 per annum, without commission on rents, and that his hours should be from ten to four.

The apothecary was paid £100 per annum for full-time work. He was allowed one free pupil and was supplied with two rooms, coal, gas and attendance. The matron was paid £80 per annum and was assigned four rooms, coal, gas and attendance. Porters each received 15 shillings per week, a free house, uniform, coal and gas. A gratuity of three guineas was paid to one and one guinea to the other. The elder porter was allowed 5 shillings a week for a boy to do his work. Of the nurses and wardsmaids, seven nurses each received £15 per annum and a gratuity with everything except tea and sugar, and five wardsmaids each received £9 under the same conditions. The cook only received £10 1s 0d, which the standing committee felt to be an insufficient amount, but, as no sum appeared to be received for refuse, it was thought that she probably had perquisites of some amount.

With the removal of the secretary's pigs and fowl, and eventual clearance of the potatoes and vegetables, it was decided by the standing committee to 'prepare plans for the pleasure grounds of the hospital'. Garden seats were provided by Dr Arthur Wynne Foot, and early photographs show that the grounds were attractive with lawns, trees and shrubs.

Lambert Ormsby, who was appointed surgeon after the death of Robert St J Mayne in 1871, showed signs of a strong character at an early age. On 27 March 1876, the joint committee called in Mr Ormsby and cautioned him for making statements which they considered disrespectful to the committee and injurious to the good working of the hospital.

In July 1878, the matron, Mrs E Jones, complained of Ormsby's rude conduct towards her. They were both called in and 'after much deliberation the Board expressed a hope that there would be better feelings in the future between the parties'.

Resignation and Death of William Stokes

William Stokes (senior) resigned in 1875 at the age of 71, having served the Meath Hospital for 49 years. He had become world famous for clinical teaching, in association with Robert Graves, and particularly for his contributions to cardiology. He was supreme among the physicians in Dublin: regius professor of physic 1840; president of the College of Physicians in 1849; fellow of the Royal Society in 1861. His election as president of the Royal Irish Academy in 1874 reflected his broad cultural interests and his knowledge of archaeology. He had received honorary degrees from Cambridge, Oxford and Edinburgh, and the Prussian Order of Merit in 1876.

In June 1876 the medical board decided to place a bust of William Stokes, together with that of Philip Crampton, in the front hall of the hospital. Stokes wrote a warm letter of thanks.

5 Merrion Square N, Dublin
30 October 1876

My Dear Ormsby,
Will you convey to the medical board of the hospital how deeply I have felt their kindness in placing my bust in the Medical Valhalla of the Meath Hospital. I can never forget it now, the long days of communion, friendship and confidence with every one of my colleagues. Accept from me my full and earnest gratitude.

Ever faithfully,
W Stokes

William Stokes died in January 1878 and was buried in St Fintan's graveyard in Howth. Fifty students of the Meath Hospital walked in procession before the coffin. The medical board passed the following resolution:

> That the medical staff of the Meath Hospital and County Dublin Infirmary take this—the earliest opportunity afforded them since the lamented death of Dr Stokes—to give expression to their deep sympathy with his family in their bereavement. They desire also to record their sense of the loss sustained by the profession of medicine at home and abroad by the death of their late friend and colleague, that honoured and beloved physician, who for a period of nearly fifty years laboured in the cause of humanity in the wards of this hospital, which owed so much to him, and upon which his many noble qualities of heart and mind—his genius, intellect, and fame shed such lustre and renown.

A year later his son, the surgeon Sir William Stokes, presented a bust of Robert Graves to the medical board of the Meath. He described it in an accompanying letter: 'It is by the late Mr Hogan [sculptor John Hogan (1800–58) was born in Cork and spent twenty-four years in Rome] and is considered to be a faithful representation of our illustrious countryman.' This bust of Graves, which is now kept in the medical boardroom, joined the busts of Stokes and Crampton, and both the Collis surgeons, in the hospital.

After the death of Stokes, his colleague Arthur Wynne Foot became senior physician to the hospital. He also was a product of the Meath and had been indentured to Maurice Henry Collis. He was a very successful teacher and was appointed professor of medicine in the College of Surgeons. The successor to the place of Stokes was John William Moore, who had been clinical clerk to both Stokes and Hudson. He had obtained a scholarship in classics at Trinity College and had graduated in medicine in 1869. He had inherited a gift for languages from his father, William Daniel Moore, a Dublin physician, and he was able to read Scandinavian. Before he was appointed to the Meath in 1875, he had been 'visiting physician' to the Cork Street Fever Hospital, and in 1871 he had passed the MD and also the newly established Diploma of Public Health at Trinity.

Barber–Bury Wing, 1887

In 1887 the Barber bequest of £4,500 enabled a large wing to be constructed to the south of the east end of the original building. The top two floors contained identical wards of 17 beds each. The upper one was medical and the one on the second floor was surgical. The first floor comprised the matron's quarters and the boardroom, and the ground floor was originally residential quarters for the registrar, or secretary. This was later to become the X-ray department. The wards were furnished from the Bury bequest and were originally known as the Bury wards. Each still has an inscription over the door commemorating the bequest. This extension made a vast difference to the accommodation in the hospital. The Bury wards were officially opened by the Lord Lieutenant and the Marchioness of Londonderry in 1888.

CHAPTER SIX

The Meath Hospital in the Confident Nineties, 1890–9

In the 1890s—apart from the laundry, the post-mortem room and the mortuary at the back, the dispensary along Long Lane and the Porter's Lodge near the gate—the hospital grounds of about three acres were still open, with lawns, shrubs and trees. However, as mentioned previously, there was a small building on the west side containing the two epidemic wards, each with 5 or 6 beds, and wooden huts had also been erected in the grounds during epidemics. The west wing for infectious cases was not built until 1896.

The annual report for 1892 contains the following statement, which we see repeated in subsequent years:

> The hospital is situated in a most healthy locality, and stands upon about three acres of ground, which affords opportunity for recreation in the open air to patients able to avail themselves of it, it is yet in close proximity to some of the poorest and most overcrowded parts of the city—including the Liberties—and is much availed of by the inhabitants of these parts. It is also very convenient to the rich townships of Rathmines and Rathgar; also Harold's Cross, Terenure, Rathfarnham, etc. Hundreds of domestic servants, labourers, and other poor from these districts, and every part of the County of Dublin are treated in its wards every year. Patients also seek admission from all parts of Ireland . . .

It is strictly undenominational. Patients are admitted without any question as to their religion, while the fullest liberty is given to the ministers of every religion to visit patients of their own persuasion.

There were 142 governors and governesses and of these the following 21 were elected, in 1892, to the standing committee (or joint committee): The Viscount Powerscourt, Powerscourt, Bray; Arthur Andrews, Newtown House, Blackrock; George F Brooks, Pickering, Celbridge; Vere Ward Brown, Balnagowan, Palmerston Park; John V Cassidy, 53 Upper Mount Street; Colonel Gerald R Dease, Celbridge; Arthur Wynne Foot, 49 Lower Leeson Street; Sir Percy R Grace, Boley, Monkstown; Sir Howard Grubb, 51 Kenilworth Square, Rathgar; Samuel E Hamilton, Grosvenor Park, Rathmines; John Hatchell, Fortfield House, Terenure; R Henry A M'Comas, Homestead, Dundrum; William M'Comas, The Grange, Kill Avenue, Monkstown; Luke John M'Donnell, 38 Merrion Square; L Hepenstal Ormsby, 92 Merrion Square; Sir George B Owens, 126 Lower Baggot Street; Alderman George Perry, 81 Harcourt Street; Sir George Porter, 3 Merrion Square; William Henry Porter, 3 Merrion Square; Judge Purcell, 71 Harcourt Street; Philip C Smyly, 4 Merrion Square.

The medical board was composed of the two physicians and six surgeons of the senior staff. The physicians were Arthur Wynne Foot and John William Moore, and the surgeons were Sir George Porter, Philip Crampton Smyly, Rawdon Macnamara II, Lambert Ormsby, William Hepburn and Sir William Stokes. They were a very distinguished group and each are the subject of a biographical sketch in Ormsby's history of the hospital.[71] The clinical assistants were James Craig and Edward Lennon, of whom we will learn more later. The secretary of the hospital was Francis Penrose, who, as was customary in those days, was resident. Penrose was to retire in 1919, after 32 years of service. The resident apothecary was Tenison Lyons, who would also eventually complete a period in office of 32 years. There was also a single resident medical officer, known as the resident surgeon, or house surgeon.

Miss Ellinor Lyons was the matron, with the official title of 'lady superintendent', and she was head of the Dublin Red Cross Order of Nursing Sisters. The ward sisters usually lived in rooms adjacent to the ward. They were known by their first names and, at that time, they were Sister Ina (Egan), medical landing, Sister Florence (Bradburne), surgical landing, later to become matron, Sister Madelaine (Orpin), children's ward, Staff Nurse Ellen (Kelly), accident landing, and Sister Kate (Walker), fever department. Sister Jessie (Irwin) was the night sister and Sister Meta (Lyons) was in charge of the convalescent home in Bray. The probationers came from the

Dublin Red Cross Sisters' House at 87 Harcourt Street, which later became the National Children's Hospital.

During the year ending 31 March 1892, the daily average number of beds occupied in the hospital was 93.6, and the daily average in the Bray convalescent home was 11.11. The total number of in-patients treated in the year was 1,521, and there were 4,484 accidents and 9,138 dispensary cases.

There were strict rules for the conduct of patients, which were probably indicative of contemporary problems, although some of these regulations might well be applied today.

1. The patients are to observe the greatest regularity and decorum at all times, are not to eat or drink of any other thing than such as are ordered by the physicians and surgeons, and supplied by the hospital; they are to be civil and amenable to the officers and nurses, and to attend to the directions they receive from them, and not to leave the wards without permission, or stand about the passages.
2. They shall not swear, use profane, abusive, or obscene language, game, quarrel, or make any offensive noise in this hospital, under pain of dismissal.
3. They shall not smoke tobacco, dirty the bedclothes, or spit on the walls or floors.
4. Complaints with respect to diet may be quietly and respectfully addressed to the secretary, Lady Superintendent, or one of the visiting governors, or medical attendants.
5. Patients able to work are to make themselves useful in any way they can, under the direction of the nurse; when able, they are to make their own beds and are to wash and clean themselves daily.
6. The clergy of all persuasions shall be at liberty to visit any patient in the hospital, on application being made to the secretary, house surgeon, or Lady Superintendent, for that purpose.[72]

Finances, 1892

The cost of running the hospital and the convalescent home for the year ending 31 March 1892 was £4,837. The hospital was largely dependent on subscriptions, donations, bequests and rents, which formed about two-thirds of the income. There was a parliamentary grant of £686 and Dublin Corporation gave £300. Being the County Dublin Infirmary, there was also a 'presentment' of £600 from the Dublin County Grand Jury. In April 1892,

contd on page 73

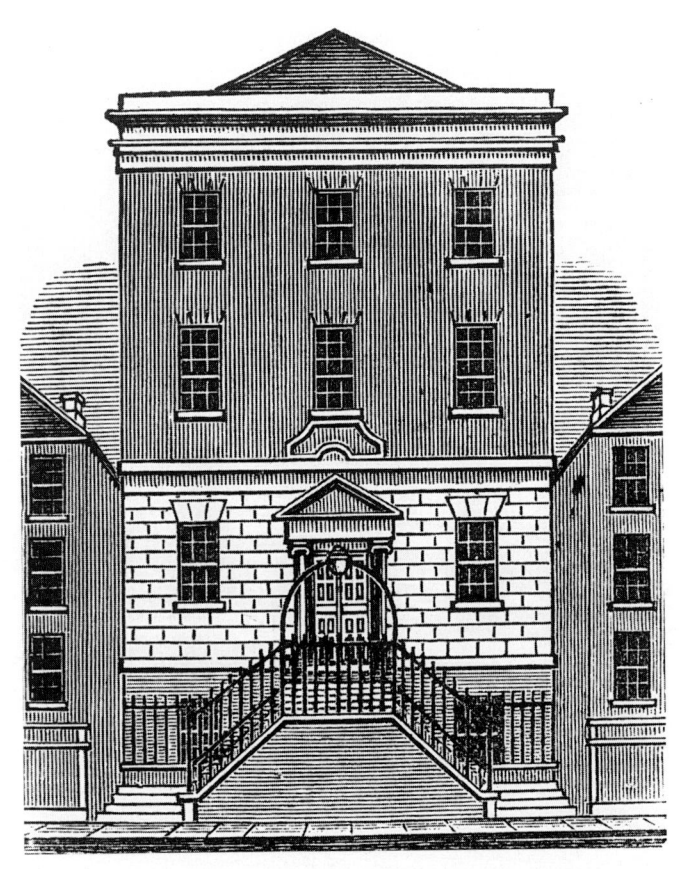

The original hospital in the Coombe, 1771.

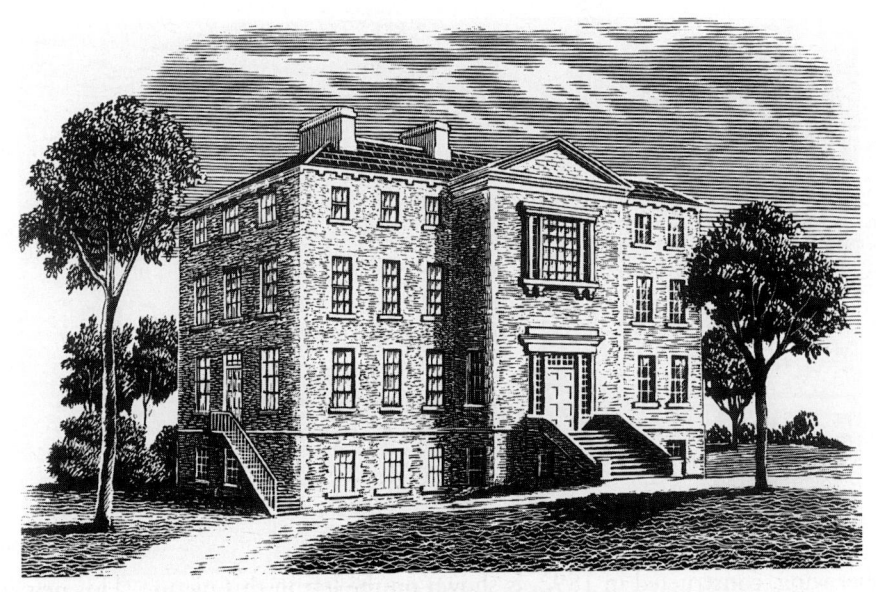

The 'new' hospital at Dean's Vineyard, Long Lane, 1822.

A drawing from 1875, showing the Smyly ward extension on the left. The two-ward fever building is seen to the right, set apart from the main hospital.

The Barber wing, constructed in 1892, is shown on the left in this picture. This new wing almost doubled the capacity of the hospital.

A view of the first nurses' home (built in 1907), as photographed in 1920.

The entrance hall in 1910. The lantern and tiling are features in memory of the Dease surgeons.

A view of the accident ward in 1900.

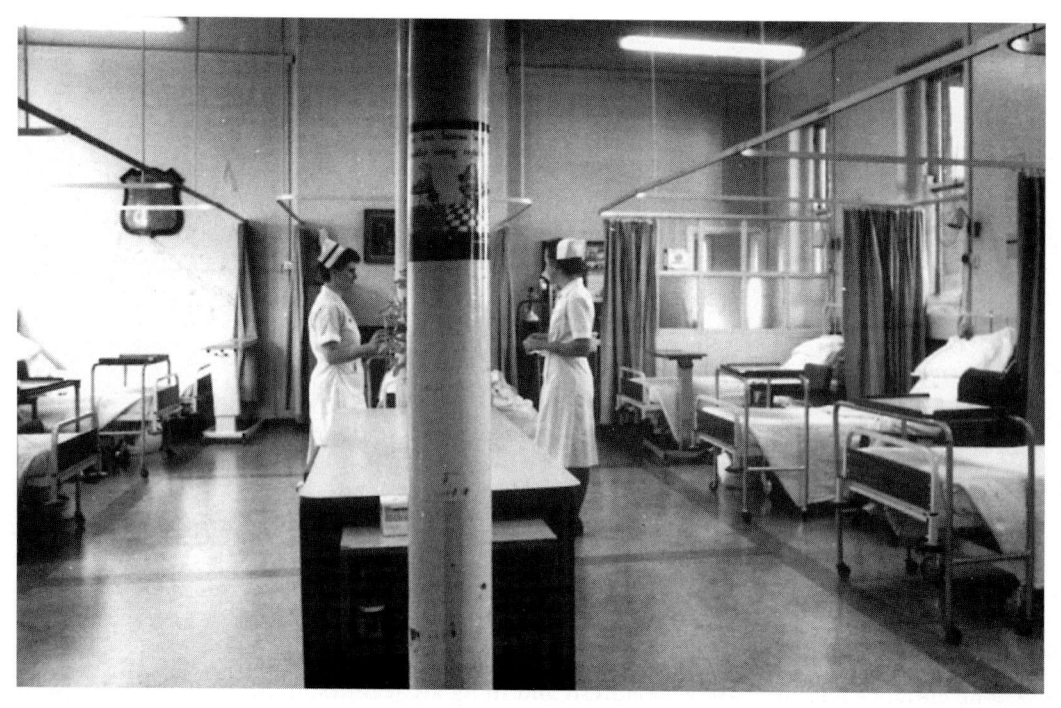

The same ward in 1995.

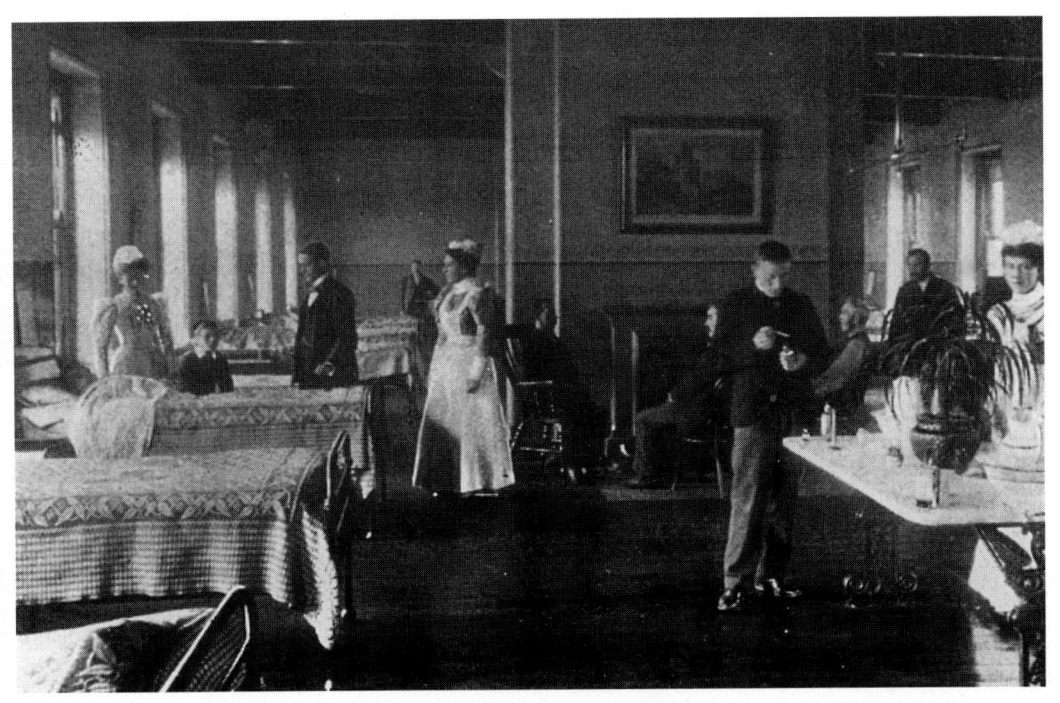

A view of ward 12 in 1900.

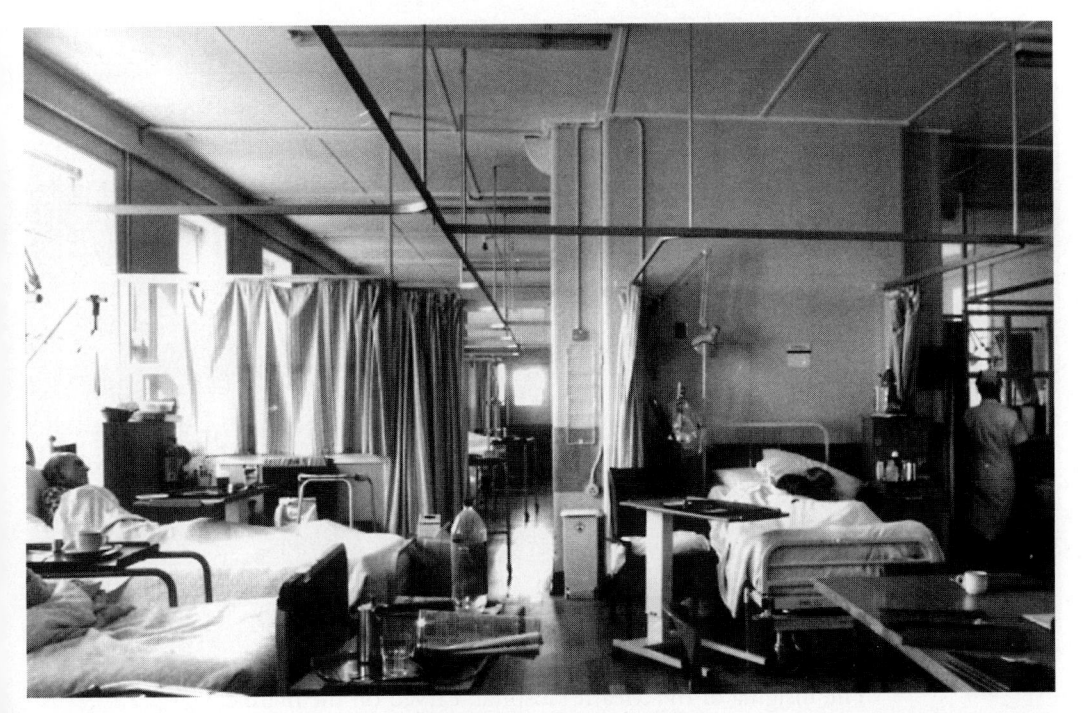

The same ward in 1996.

Lambert Ormsby's ornamental main staircase and rooflight, constructed in 1897.

The insignia of the Meath Hospital, carved on the marble mantelpiece in the entrance hall.

David McBride, one of the original surgeons and founders of the hospital in 1753.

Sir Philip Crampton, surgeon 1798-1858.

William Henry Porter, surgeon 1819-1861.

Robert J Graves, physician
1821-1843. (Courtesy of RCPI)

William Stokes, physician
1826-1875. (Courtesy of TCD)

CLINICAL REPORTS,

OF THE

MEDICAL CASES

IN THE

MEATH HOSPITAL

AND

COUNTY OF DUBLIN INFIRMARY,

DURING THE SESSION 1826, 1827.

PART I.

BY R. J. GRAVES, M. D.

AND

WILLIAM STOKES, M. D.

PHYSICIANS TO THE HOSPITAL.

DUBLIN:
PRINTED FOR THE AUTHORS.

TO

JOHN CHEYNE, M. D. F. R. S. E.

PHYSICIAN-GENERAL,

ORMERLY PHYSICIAN TO THE MEATH HOSPITAL,
&c. &c. &c.

THIS WORK,

OF WHICH THE FOLLOWING PAGES ARE THE

COMMENCEMENT,

IS DEDICATED,

AS A MARK OF RESPECT FOR TALENTS,

ASSIDUOUSLY AND SUCCESSFULLY EMPLOYED

IN THE IMPROVEMENT OF

PATHOLOGY,

BY HIS OBEDIENT SERVANTS,

R. J. GRAVES AND W. STOKES.

Clinical reports; title pages of work by Graves and Stokes in 1827.

RULES

FOR REGULATING THE DUTIES OF

APPRENTICES.

I.—Each Apprentice, during his attendance on the Hospital, shall behave himself with decency, regularity, and the utmost tenderness and humanity to the Patients, as well as with the greatest respect and obedience to the Physicians and Surgeons, and civility to all the other officers of the Hospital; and no Apprentice shall waste any of the medicines, ointments, or lint, nor use them for any other purpose but the regular and necessary purposes of the Hospital, and only within its walls. Nor shall they, or any of them, write any prescription, unless such prescription shall be directed by the Physician or Surgeon in attendance; and every Apprentice transgressing any part of this rule, shall be subject to dismissal from the Hospital.

II.—No Apprentice or Pupil to be admitted into any of the Wards, but in the presence of one of the Physicians or Surgeons, except during the regular hours of attendance, or in cases of urgency, when he shall first produce to the House-keeper, or Registrar, a written order from the attending Physician or Surgeon.

III.—No Pupil or Apprentice shall remain in the Hospital after the regular time of daily attendance is over.

IV.—The attending Surgeon shall appoint an Apprentice, upon whose character for respectability, regularity, good morals, and competent professional knowledge he can rely, and who shall have served at least three years of his apprenticeship to be "Resident Pupil" for six months, whose name shall be reported to the Standing Committee immediately after his appointment. He shall be always in attendance during the day, except at his meal times; he shall sleep in the Hospital, and shall not absent himself, without leaving a memorandum in writing, with the Apothecary, stating the time he will return, and where he may be found in the interim. The Surgeon of the month to be responsible for the orderly and correct conduct of such Pupil.

Rules for apprentices, c.1830.

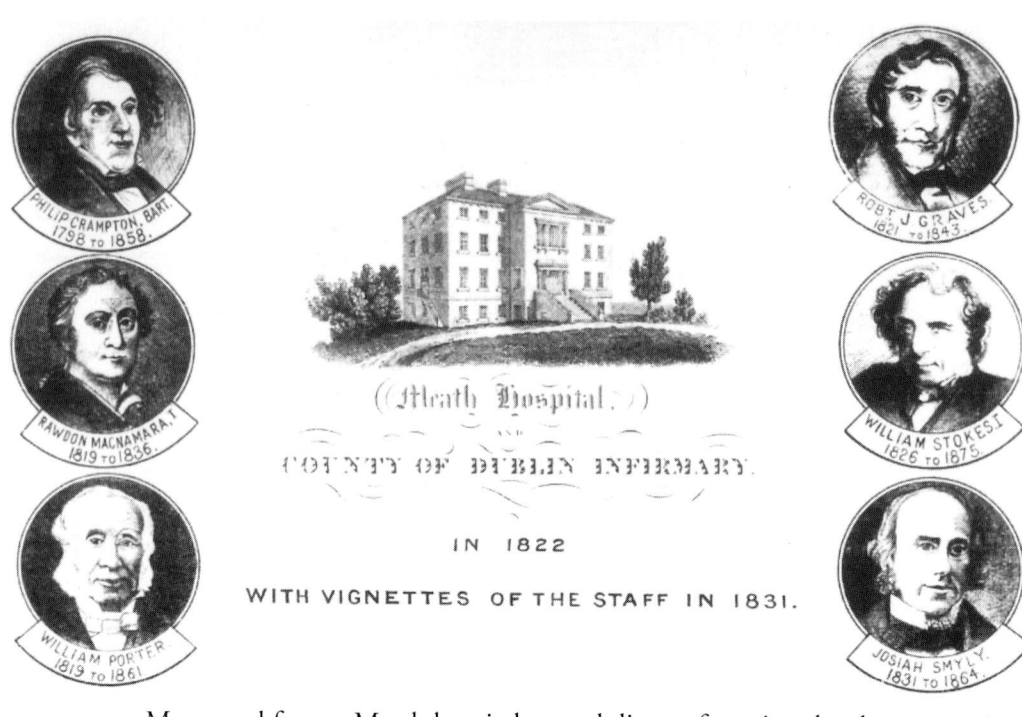

Menu card from a Meath hospital annual dinner, featuring sketches of the staff from 1831.

Francis Rynd, surgeon 1836-1861.

Josiah Smyly, surgeon 1831-1864.

Sir George Hornidge Porter, surgeon
1849-1895.

Maurice H Collis, surgeon
1851-1869.

Sir Philip Crampton Smyly, surgeon
1861-1904.

Rawdon Macnamara II, surgeon
1861-1893.

Sir William Stokes, surgeon 1864-1868 and 1888-1900.

Sir Lambert Ormsby, surgeon 1872-1923.

William J Hepburn, surgeon 1879-1911.

CHILDREN'S WARD.

Meath Hospital & Co. Dublin Infirmary.

A BAZAAR AND SALE OF WORK

Will be held (D.V.), weather permitting,

At BUSHY PARK, Terenure,

THE BEAUTIFUL DEMESNE OF

SIR ROBERT SHAW, Bart.,

(KINDLY LENT FOR THE OCCASION),

On TUESDAY, the 27th of MAY, 1884,

IN AID OF THE

CHILDREN'S WARD, MEATH HOSPITAL.

The Sale will be held in a **GIPSY CAMP**, and the Ladies presiding at the Stalls will be dressed in **Fancy Costumes**.

The Bazaar will be open from 2 till 8 o'clock in the Evening.

ADMISSION, SIXPENCE.

N.B.—The Rathfarnham Tram Cars pass Bushy Park Gate.

Contributions of Work and other Articles may be sent not later than Monday, 26th May, 1884, directed to Miss BAYLEE, *Hon. Sec.*, Newcourt, Terenure.

Advertisment for a bazaar to raise funds for the Children's Ward in 1884.

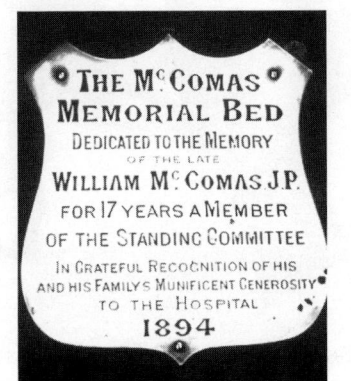

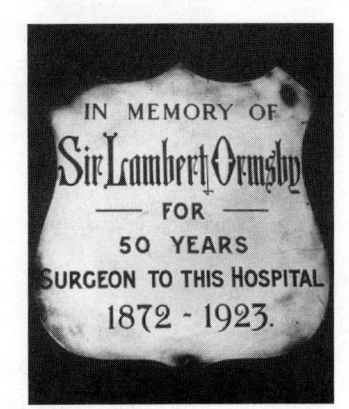

Plaques from the Victorian period, commemorating the endowment of beds. This was a vital source of income to the hospital.

Junior medical staff in 1887.
(Left to right) Standing: W Wynne, J A Burland, J Ryan,
A McFarland, H Hildige.
Sitting: Edward Lennon, senior assistant physician; F P Newell,
house surgeon; James Craig, junior assistant physician.

Meath Hospital Red Cross Sisters in 1892.
Front: Sister Florence (surgical, later matron), Sister Ellen (accident)
Centre: Sister Ellinor (lady superintendent)
Back: Sister Ina (medical), Sister Kate (fever)

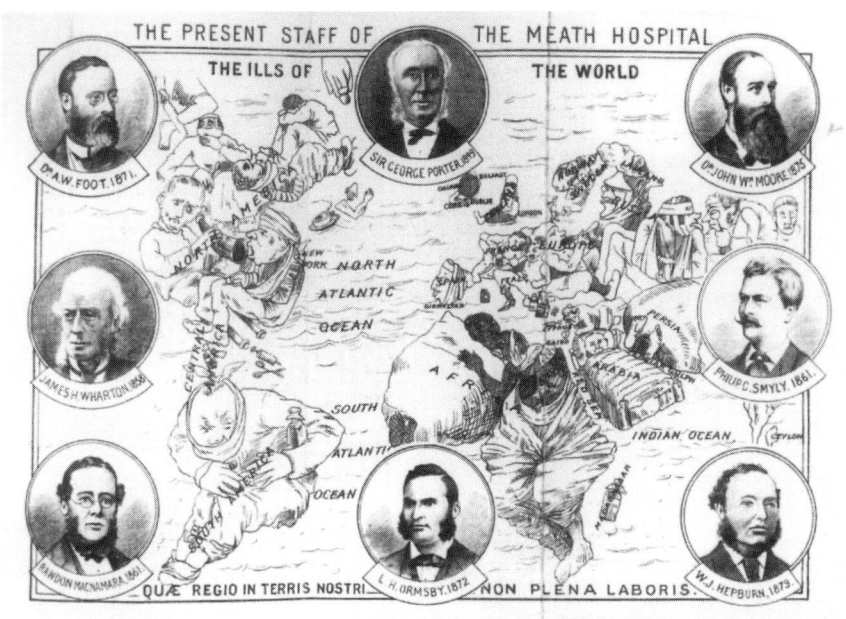

DUBLIN'S MEATH HOSPITAL

A menu card from the annual dinner of 1887.

MEATH HOSPITAL and CO. DUBLIN INFIRMARY.
Programme of Clinical Teaching—Session 1897-98.

	Monday.	Tuesday.	Wednesday.	Thursday.	Friday.	Saturday.
CLINICAL MEDICINE. (Throughout the Session.)		9 a.m. Dr. J.W. Moore		9 a.m. Dr. J. Craig		9 a.m. Dr E.E. Lennon
CLINICAL SURGERY.	9 a.m.		9 a.m.		9 a.m.	
1897. OCTOBER.	Sir W. Stokes		Mr. Patteson		Mr. Hepburn	
NOVEMBER.	Sir W. Stokes		Sir P. C. Smyly		Mr. Patteson	
DECEMBER.	Mr. Ormsby		Mr. Ormsby		Mr. Ormsby	
1898. JANUARY.	Mr. Patteson		Sir P. C. Smyly		Mr. Hepburn	
FEBRUARY.	Sir W. Stokes		Sir P. C. Smyly		Mr. Hepburn	
MARCH.	Sir W. Stokes		Sir P. C. Smyly		Mr. Patteson	
APRIL.	Sir W. Stokes		Mr. Patteson		Mr. Hepburn	
MAY.	Mr. Ormsby		Mr. Ormsby		Mr. Ormsby	
JUNE.	Mr. Patteson		Sir P. C. Smyly		Mr. Hepburn	

Operations admitting of delay will be performed at 10 o'clock a.m., on Mondays, Wednesdays, and Fridays, by the respective Surgeons on duty.

Timetable drawn up to facilitate teaching of surgery, 1897.

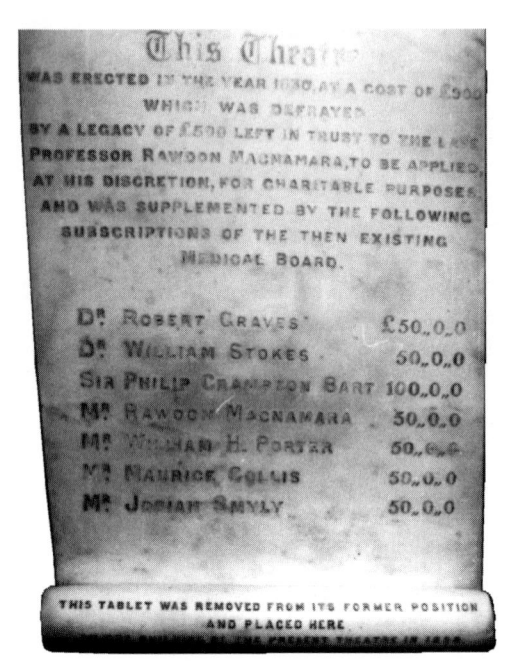

Marble tablet marking subscriptions to the first operating theatre in 1830. This tablet was moved in 1898 and placed near the first aseptic operating theatre.

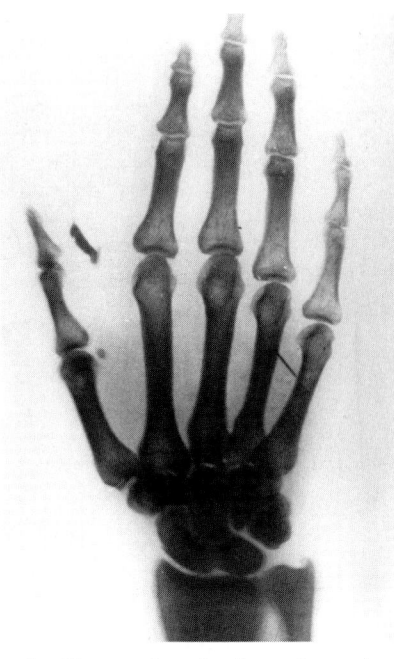

An X-ray taken by Lane Joynt in 1897, two years after Rontgen's discovery. A needle can clearly be seen in the hand.

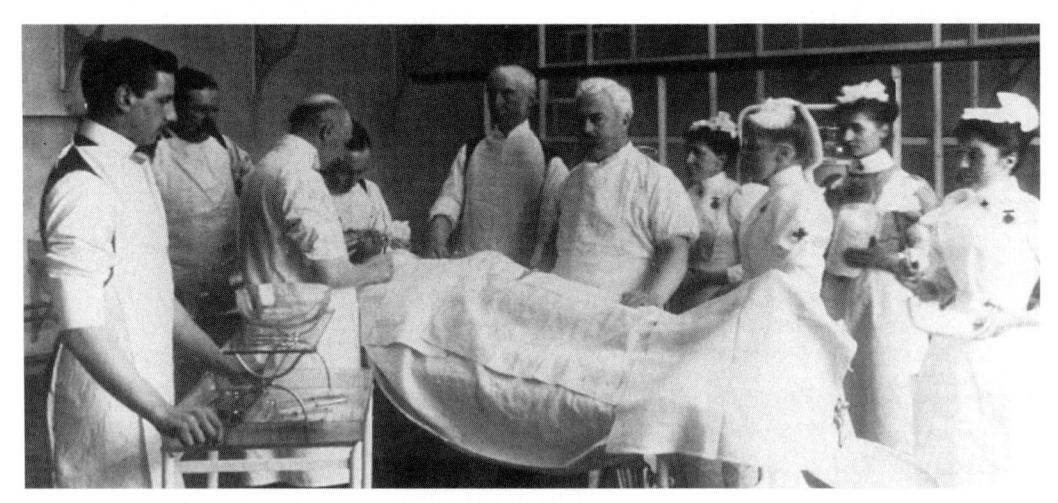

First Aseptic Operating Theatre, 1898.
Second from left: Tenison Lyons, apothecary
Fifth from left: Sir William Stokes
Sixth from left: Sir Philip Crampton Smyly
At patient's feet: Ellinor Lyons, matron
Extreme right: Sister Florence (Bradburne), later matron

contd from page 56

this was increased to £1,000, as a result of a deputation from the hospital led by Sir Lambert Ormsby.

The main expenditure was on 'provisions', and this amounted to £1,706 in that year. About £800 was spent on salaries, which mainly went to nursing staff and servants. It was noted in the joint committee minutes of February 1891 that the salary of Sister Ina of the medical landing was to be increased to £30 per annum, but this was considered 'exceptional', in view of her seniority.

The physicians and surgeons gave their services free to the hospital, which was primarily intended for the care of the poor. They made their living largely from attending the rich in their homes. Teaching fees from medical students were significant. The medical board minutes in 1905 noted that the total fees collected for the summer session amounted to £163 6s 11d, giving each of the seven members of the board £23 6s 8d (thanks being recorded to 'Mr Hepburn for waiving his share').

There is a suggestion that the surgeon William Joseph Hepburn did not have the same position as his colleagues. Ormsby records in his history that Hepburn, who was well qualified:

> was appointed surgeon in 1879 in place of Robert Pearse White deceased. This was related to the will of John Bury, Esq, BL, sometime residing at 110 Leinster Road, Dublin, possessing large estates in counties of Meath and Kildare, died, leaving his entire property, when realised, to any Dublin Hospital, which would build a wing or ward for medical and surgical cases in memory of the testator and that this then present medical attendant, William Joseph Hepburn, should be medical officer of such wing or ward.[73]

Hepburn's appointment was therefore tied to the Bury legacy to the hospital and, as such, might be regarded as a 'purchase appointment'.

Standing Committee

The elected group of governors and governesses which managed the hospital was originally known as the 'standing committee' and, from about 1900, as the 'joint committee', and from 1961 it became the 'hospital board'. During the 1890s the standing committee usually dealt with tenders for supplies, with legacies, with recording thanks to donors and replying to subscribers and others about procedures for admission of patients. Appointments to the medical staff were made by the medical board and this appeared at that time to be accepted without question by the standing committee, although in May 1892 the latter showed some authority by demanding three names from the medical board to be submitted for the election of house surgeon.

The standing committee was also concerned with such domestic matters as the hydraulic lift. This was for conveying goods only, as a proposal at that time for a lift for patients had been 'abandoned'. The hydraulic lift depended on an ample supply of water and was in need of a new 'ram'. Later, in 1893, it was noted that water only reached the roof cisterns at night, because Vartry water was being diverted to the canal, and that there was a major problem with the hospital drains which would have to be 'relaid to Heytesbury St instead of Long Lane'. Destruction of typhoid stools was also discussed and a 'cremator' proposed.

West Wing, 1896
The accommodation in the 10-bed building for infectious fevers had always been inadequate, and excess patients had been accommodated in temporary huts during severe epidemics, as already mentioned. On 12 March 1894 a sub-committee appointed by the joint committee to report on a project to construct a new 'west wing' stated that Dr Edward Emmanuel Lennon, who had been appointed surgeon in place of the late Rawdon Macnamara in May 1893, had offered financial support for the project:

> relative to the application of the Hughes Bequest, I beg leave to say that, I will hand it all over to the Standing Committee of the Meath Hospital, towards the erection of the proposed Epidemic Wards, on condition that the Standing Committee appoint me Physician to the Extern Department.[74]

The standing committee referred this to the medical board, who replied that on 26 April Dr J W Moore had proposed that 'Dr Lennon be transferred from surgical to the medical staff with the title and full standing of physician to the hospital, and that Sir George Porter, chairman, refused to receive the motion on the grounds of its illegality.' However, Dr Lennon did become physician in 1893 and the sum of £1,934 from the Hughes bequest was paid towards the construction of the west wing, the total cost of which was finally £5,506.

The west wing was opened by Lord Powerscourt on 6 October 1896, and later that day the Lady Superintendent was permitted to host an 'At Home' with music and dancing, 'provided it ended at 9.00 pm'. It was resolved by the standing committee that 'none but infectious cases be admitted into the West Wing'. It was agreed to name the two lower wards after Hughes, and brass plates with the following inscription were erected over the entrance of each ward:

> This ward is dedicated to the memory of Mrs Ann Hughes of Marlborough Road in the County of Dublin who died 9 March 1893 and is endowed with money which she left to be used for the furtherance of such hospital work as her executor Edward Emmanuel Lennon, one of the physicians to this hospital, should select. In the exercise of his trust he has handed the bequest over to the governors of the Meath Hospital 1897.

Edward Emmanuel Lennon

Edward Lennon was a tall handsome man who remained a very eligible bachelor for many years. It was said that when he entered the wards the nurses' hearts were 'all of a flutter'. He married late and had two children, neither of whom entered the medical profession. He had a reputation for making a quick spot diagnosis and also for having a talent for raising money.[75] Boxwell describes him in his 'Historical Sketch of the Meath Hospital and County Dublin Infirmary':

> Lennon was the son of a well-known Irish landowner in Co Meath, and one of a large family. Brought up in the belief that there would be plenty of money for all, he was not educated for any career in particular. But finding *res angusta domi* on his father's death, he set out to seek his fortune in America. He worked his passage out and, in the USA, he learnt various trades including cobbling shoes. He made half a dollar a day as a stone cutter, and worked as a 'wracker' in a drinking saloon. Finally, having earned enough money trundling a wheelbarrow in Armour's canning factory in Chicago to bring him home, he returned to Dublin and decided to become a doctor. He had a brilliant medical course in the Royal College of Surgeons' School and at the Meath, was elected fellow of the Royal College of Physicians in Ireland in 1892, and physician on the hospital staff in 1893. He soon developed a very large practice and was one of the best teachers of clinical medicine ever at the Meath.[76]

Improvements at the Meath

In 1895 Lambert Ormsby was also responsible for obtaining money for the hospital for structural improvement. Although the amount was 'not less than

£250', and minor compared with that for the west wing from the Hughes bequest through Lennon, it was directed to an admirably aesthetic cause. This was to rebuild the staircase in the main building with an ornamental railing and to construct a decorated ceiling and rooflight. The principal donor was Michael J Kenny and this is recorded by a brass plaque on the first half-landing. Also in 1895, the joint committee agreed to pay £290 towards building the pathological department, and, in 1897, accepted a tender of £2,050 for building a new theatre, accident ward and laundry. The new theatre was described as the 'first aseptic theatre'.

In 1898 it was resolved to rent new premises for the convalescent home in Bray for ten years, at £45 per annum. This was Abbey View, two terraced houses on the Dublin Road. Telephones were introduced for the medical staff in 1899, but the cost was to be shared with the hospital. There was still no electricity in the hospital, which was lit by gas.

The Local Government Act of 1898

At a time when Home Rule was becoming a real possibility, the Local Government Act of 1898 was passed, establishing elected County Councils in Ireland and 'thus taking the administration of local affairs out of the hands of the property-qualified grand juries and the local gentry'.[77] Under the 1898 Act, the Local Government Board of Ireland, in a letter to the hospital on 19 April 1899, fixed the number of the 'Joint Committee of Management of the County Dublin Infirmary at 20, of whom 3 are to be appointed by the County Council, and 17 by the Governors and Governesses of the Infirmary. This arrangement will take place on 22 May 1899.' The joint committee replied that the number designated by Act of Parliament was 21, and so resolved, on 31 July 1899, that the composition would be 17 representing the governors and governesses, 1 to represent the City Corporation and 3 the Dublin County Council. The three councillors appointed by the County Council in May 1899 were J J O'Reilly of Cabinteely, Captain Vesey of Lucan House, and James Mahoney of 7 Raglan Road.

When the Local Government Act was being considered by the government, members of the medical board were concerned that their rights of appointing their successors might be affected. John Moore and James Craig had an informal interview with the attorney-general, the Rt Hon John Atkinson, on the subject of this privilege. However, after receiving correspondence from Horace Plunkett, MP, which included the following extracts of the debate in parliament, it was resolved by the medical board not to attempt any amendment to the Local Government Bill.

On the 24 May 1898 in the House of Commons, Horace Plunkett had moved, 'Nothing in this Act contained shall affect the mode of election of the physicians and surgeons of the Meath Hospital and County Dublin Infirmary who attend and serve same without fee, salary or reward.'

Mr Dillon (Mayo E)—'I would be in favour of leaving the medical staff to elect their colleagues but to state deliberately that they attend without fee, salary or reward is simply ridiculous, because I know from experience they do.'

Mr Clancy (Dublin S)—'I am rather surprised the Government will not consent to the suggestion thrown out. I should not be surprised because the amendment is meant to preserve the Meath Hospital which has been a tax on the people of Dublin for the last fifty years. Although it gets £1,000 a year from the County of Dublin, where the large majority of the ratepayers are Catholics, and gets £300 a year from the city—this even to the present year—out of 21 Governors only 3 are Catholics.'[78]

Chapter Seven

Queen Victoria's Visit and the Appointments Controversy, 1900–12

And so, by the turn of the century, the composition of the joint committee was legally confirmed. The hospital had been greatly improved since August 1897 by the separation of infectious patients in the new west wing. In September 1899, 37 cases of fever were accommodated there, mostly of typhoid and measles. Surgery was practised in improved conditions with antiseptic precautions. Anaesthesia was administered with chloroform or ether, although specific anaesthetists were not yet appointed. Nursing training had been well established in conjunction with the Red Cross Nursing Home in Harcourt Street. The first X-rays had been taken by the newly appointed clinical assistant, Richard Lane Joynt, in early 1897.

Queen Victoria's Visit
In April 1900, at the age of 81, Queen Victoria visited Dublin for three weeks. The Boer War was in progress at the time and many Irishmen were serving there in the British forces. Her visit was thought by some to be partly a gesture of recognition towards their contribution to the Empire. At the time in Dublin there was a noisy Irish minority supporting the Boer cause. The joint committee, in 1899, had agreed it was 'willing to accept sick and wounded Irish soldiers from the seat of war in South Africa at a minimum charge of three shillings per day'.

During her stay at the Viceregal Lodge the Queen, or members of her entourage, visited various institutions in Dublin, such as the Royal Hospital Kilmainham, the Adelaide, St Vincent's Hospital, the Royal City of Dublin Hospital, Baggot Street, the Mater Hospital, and the Convent of the Sacred Heart, Mt Anville. On 20 April 1900, the Queen came in her carriage to the Meath, taking a route from Phoenix Park via Rialto and the South Circular Road. The minutes of the joint committee recorded the event:

> The secretary reported that the Queen accompanied by Princess Christian and the Princess Henry of Battenberg visited the Hospital on Friday afternoon the 20th April at 4.30 o'c. He, the secretary, received an intimation from the Earl of Denbigh by telephone at 1.30 pm of the visit, the notice being so short he was unable to notify many of the members of the Committee in time to be present. The hospital was gaily decorated with flags and trophies, and a triumphant arch was erected at the entrance gate. Her Majesty's carriage drove to the front steps on which the nursing staff was grouped, the patients being grouped opposite at the dispensary . . .
>
> The nurses and patients were pointed out to the Queen who turned to each group bowed and smiled. Lord Denbigh mentioned to Her Majesty that the notice was very short. The Queen said that everything was very nice indeed. I am much pleased she added that the position of the hospital is very good. Both on Her Majesty's arrival and departure she was very heartily cheered.[79]

The report to the joint committee of the visit to the hospital and the enthusiastic welcome was somewhat offset by the rather miserly complaint later at the cost of the decorations supplied for the Queen's visit. The bill from Millar and Beatty for £39 4s 0d for supplying flags was considered excessive by the committee, and it was decided to send a cheque for £15 'in full discharge'. Later the bill was reduced to £31.

Following the Queen's visit, the senior physician, John Moore, president of the Royal College of Physicians of Ireland, received a knighthood. The following was also noted in the minutes of the joint committee:

> The Keeper of the Privy Purse presents his compliments to the secretary and has received the Queen's commands to forward the accompanying Print of Her Majesty to the Meath Hospital and County Dublin Infirmary in recollection of her recent visit.

Changes in Medical Staff

A number of important changes occurred in the honorary staff in 1900. William Taylor, who had been clinical assistant for two years, was appointed visiting surgeon in place of Robert Glasgow Patteson, who had died after only five years on the staff. William Taylor was an energetic and outspoken figure who was to have a very distinguished career, as has been graphically described by J B Lyons.[80] He was president of the College of Surgeons from 1916 to 1918 and gave outstanding service during the war. He remained on the staff of the Meath until 1922, when he transferred to Sir Patrick Dun's Hospital, an appointment which went with his election as regius professor of surgery in Trinity College.

In 1900, at the age of 61 Sir William Stokes died of typhoid fever, at Pietermaritzburg in Natal, while on duty as consulting surgeon to Her Majesty's forces in South Africa. Son of the famous William Stokes, he had been president of the RCSI in 1886 and professor of surgery since 1872. The medical board elected Richard Lane Joynt surgeon, in place of Sir William Stokes, in October 1900. He had been clinical assistant, was already pioneering radiology in the hospital and was engaged in producing 'skiagraphs' by X-ray.

In 1903 the first clinical specialist was appointed. This was Dr Kidd, who was elected as a 'gynaecological assistant'. The following rules governing this appointment were made by the medical board and approved by the joint committee.

1. That the style of the appointment be that of 'Gynaecological Assistant'.
2. That the appointment be an annual one, the holder being eligible for re-election.
3. That the holder shall have a dispensary twice a week for diseases peculiar to women and that he be punctual in his attendance.
4. That this appointment confers no right on the holder to have any beds in the hospital, or to treat any cases in the main building of the hospital.
5. That it be distinctly understood that this appointment gives the holder no claim to election on the full staff in case of vacancy other than he might have possessed had he not held such an appointment.

On the medical side, James Craig, who had succeeded Arthur Wynne Foot in 1892, moved to Dun's Hospital on becoming 'king's professor of the practice of medicine' in Trinity in 1910. The vacancy thus created for a physician at the Meath led to a serious split of the medical board over the election of a successor, and to a confrontation between the joint committee and the medical board over the power of appointment which lasted for at least two years.

The Appointments Controversy, 1910

On 29 September 1910 the medical board, by a majority vote, elected Charles H G Ross as visiting physician in place of James Craig. Ross had been house surgeon at the Meath and was medically qualified for only two or three years. The other two candidates were both fellows of the Royal College of Physicians of Ireland and much senior to Ross. They were Francis C Purser, assistant physician at the Richmond Hospital for several years, and William Boxwell, clinical assistant and assistant pathologist at the Meath for six years. Sir Lambert Ormsby, Dr Edward Lennon and Mr Lane Joynt had voted for Ross, and Sir John Moore and Mr William Taylor for Boxwell. Conway Dwyer abstained.

The joint committee were not satisfied with the decision of the medical board and requested reconsideration of the appointment. Sir John Moore, who was on the joint committee, questioned the legality of Ross's qualification, as he was a graduate of the Royal University, and the matter was referred for legal opinion. Charles L Matheson, KC, in a long review of the Acts governing the management of the Meath Hospital, gave his opinion to the joint committee on 14 November that the election of Ross was invalid. He stated as follows:

> The Meath Hospital Act of 1815 puts that institution on a different basis from other County Infirmaries in Ireland as the Physicians and Surgeons are bound to act without a fee or reward and apparently, in consideration of their doing so, they are given the privilege of electing persons to fill any vacancies in their body—Section 7 of the Act which deals with these matters is not well worded, but I have no doubt whatever that its operation and effect was to give the right and privilege of filling up from time to time vacancies that might occur in the Medical Staff not only to the then present physicians and surgeons of the hospital but to their successors, the Physicians and Surgeons of the hospital for the time being when vacancies should occur.
>
> The same section 7 expressly limits the class of persons who could be appointed to the position of physician or surgeon to 'Members and Licentiates of the King's and Queen's College of Physicians and Royal College of Surgeons'—and this section as far as I can find has never been repealed.

Naturally the medical board were pleased with this legal opinion which supported their claim to the right of appointment of medical staff. However, they believed that the necessity of limiting appointees to those holding membership or licence of the Colleges of Physicians or Surgeons, as stated under the 1815 Act, should no longer be applied and they claimed that there had been many exceptions to this in the past.

Meanwhile, Charles Ross sat the examination for the licence of the College of Physicians on 6 December 1910, and a letter was received by the joint committee on 12 December from the medical board stating that 'Dr Ross LRCPI has this day been duly and legally elected by the medical board as Visiting Physician to the Hospital in the room of Dr Craig, resigned.' This election was again opposed by Sir John Moore and William Taylor (Lambert Ormsby, Edward Lennon, Lane Joynt and Conway Dwyer voted in favour of Ross). At the same meeting, the joint committee also received a letter from the Quaker George Newson Jacob (1854–1942), a lay member who was unable to be present;

> ... as a member taking a deep interest in the welfare of the Meath Hospital, I write these few lines to advocate a different policy being adopted in the appointment of Medical Officers. The Medical Board, relying upon an Ancient Act of Parliament, no doubt have the technical right to ignore the wishes of the Joint Committee, but by doing so they are hardly acting in accord with the ideas of the present day. The committee is responsible to the subscribers and public for the efficient working and reputation of the hospital, and consequently are fairly entitled to have their views considered in the selection of Medical Officers. I am afraid that too rigid an insistence of their legal rights on the part of the Medical Board will be calculated to seriously injure the hospital.

William Taylor, who as honorary secretary to the medical board was obliged to communicate to the joint committee that Ross had been elected, nevertheless violently disagreed with the decision. He also submitted a letter, which was six pages long, exposing, as he saw it, the evil motivations of the majority of the members of the medical board. In this lengthy protest, he pointed out that Ross was nephew of Edward Lennon and that, since he had left the Meath Hospital as house surgeon, Ross had 'spent the greater part of the succeeding year travelling with a patient round the world. Early this year he started practice as a general practitioner in Navan'. He wrote that Dr Lennon gave as his reason for voting for Dr Ross his desire to break up the formation of a 'family party', which he said was about to be established if Dr Boxwell was elected and if his cousin, Henry Stokes, now assistant surgeon, later became surgeon. 'He omitted to inform those to whom he gave this reason that by voting as he did he was merely establishing another family party.' Taylor also wrote that Lambert Ormsby and Lane Joynt had opposed Boxwell being appointed because he had voted against Ormsby in the past. William Taylor's long letter apparently impressed the joint committee and it was ordered that it be affixed in the official minutes of the meeting.[81]

At the next meeting of the joint committee, in January 1911, a further communication came from members of the medical board—signed by Lambert Ormsby, Edward Lennon, Lane Joynt and Conway Dwyer—protesting against the 'absurd tirade of Mr William Taylor', but they did not answer any of the specific charges made. J Mulhall recommended that a public enquiry should be held by the Local Government board. The joint committee finally resolved that 'the Medical Board be asked to consent to all medical and surgical appointments in the future being made subject to ratification by the Joint Committee. This arrangement to continue until the existing special rights of the Medical Board in the matter expire.'

At the February meeting of the joint committee, it was noted that the medical board agreed to meet with the joint committee to discuss medical appointments in the future. It was also noted that 'the Medical Board elected unanimously Dr Oliver Gogarty to fill the vacancy created by the resignation of Mr Conway Dwyer'. The committee resolved to meet with the medical board on 6 March, and also to postpone consideration of the election of Dr Gogarty. Despite this, on 10 April the joint committee received the following resolution from Mr Oliver Gogarty, who was now honorary secretary of the medical board:

> The Medical Board are anxious to meet the view of the Joint Committee at all times, and are ready to give them the most careful consideration. But the Medical Board are not prepared to waive their undoubted and legal rights in reference to election of their colleagues to the medical staff of the hospital.

In July 1911, William Joseph Hepburn died, having served as surgeon for 32 years, and William Pearson was elected by the medical board in his place. This was noted by the joint committee at its meeting in September without comment. Again Mr Mulhall expressed dissatisfaction with the action of the majority of the medical board in connection with recent elections and with the serious friction between members of the medical board. He was still of the opinion that the Local Government should be requested to hold a sworn enquiry into the situation. This was seconded by Mr Jacob. This proposal may have had some effect, because a notification to the joint committee from

Mr Pearson, honorary secretary of the medical board, in October stated that, 'On 28 September 1911, William Boxwell was unanimously elected physician to the hospital in place of Dr C Homan Ross, resigned.'

Although the immediate issue of this appointment was resolved, the correct procedure for legally appointing medical officers for the future remained controversial. The joint committee again sought counsel's opinion and in November 1911 Charles L Matheson, KC, advised that there were only two ways in which the privilege of election of physicians or surgeons possessed by the medical board 'can be got rid of'. The first and simplest is to induce the three remaining members of the medical staff who enjoy the privilege to release their privilege in favour of the committee (the three were those who had been appointed before the Local Government Act of 1898—Moore, Ormsby and Lennon). The second is to obtain an Act of Parliament giving the power of appointment of physicians and surgeons solely to the joint committee.

The medical board held a special meeting on 23 November 1911 and sent the joint committee a long statement, in which it appears they had sought legal opinion:

> The Medical Board are more than surprised that the Joint Committee, or certain members of that committee, are endeavouring to disturb and annul an arrangement which has existed since the foundation of the hospital in 1753, for the advantage of the institution, viz: 'The privilege of the remaining members of the staff to fill up vacancies on the staff as they may occur from death, resignation, or otherwise', and which privilege has been conferred on each member of the staff at his election . . . And the existing members of the staff are furthermore advised that so long as they remain on the staff they are bound individually and collectively to exercise such privilege as aforesaid, for the following reasons:
>
> By a document prepared and presented to the Irish House of Commons by the Governors of the Meath Hospital in 1773, showing that the surgeons and physicians at that time had been put to considerable personal expense in maintaining the hospital, and that also they by infinite industry and application had been the principal agents in providing funds for the erection of the hospital buildings.
>
> The petition or memorial of 1773 further states that from the first, the surgeons and physicians had served the hospital without fee or reward, and on the hospital being placed on a somewhat more permanent footing by being constituted the County Dublin Infirmary, the Medical and

Surgical Staff deliberately gave up £100 (Irish Currency) per annum (Treasury Grant) to which they were entitled. For this annual concession on the part of the medical board it was ordained and solemnly agreed to, that the Medical and Surgical staff for the time being, and their successors, should always and for all time have the power to appoint their successors on the staff as vacancies occurred . . .

By allowing the medical board to elect their successors the hospital funds benefit to the amount of £100 per year as before stated.

If the privilege is abolished the medical and surgical staff would naturally claim successfully this amount, and the hospital funds would be deprived of the same amount accordingly.

The Medical Board see no advantage whatever in making a change in the mode of election of the staff, as they are at a loss to understand how the lay members of the Conjoint Committee could make a better selection than the medical and surgical staff do at present.

The Medical Board would respectfully direct the Conjoint Committee to look back to the long list of names of those who have served on the staff of the hospital from its foundation to the present, and say whether the present system and mode of election has failed to secure the most competent and most distinguished men of the day, and such were elected irrespective of creed, politics, birth, or social position. Competence and capability were the deciding reasons for their selection . . .

The Medical Board feel that all who have the interests of the hospital at heart should unite in avoiding any legal proceedings which would involve the hospital in expense.

The Medical Board wish respectfully to intimate to the Conjoint Committee that they desire to give them due notice that they will oppose by every legitimate means in their power the proposal to deprive them of their just rights and privileges as aforesaid.

A special meeting of the joint committee was called for 22 January 1912 to consider further the question of appointment of medical staff. All members of the committee had received copies of the counsel's opinion and the reply from the medical board. After a prolonged discussion it was finally decided that Mr William S Collis, the chairman, should meet with the medical board to place before them the views of the committee. Mr Collis reported back to the committee in February that he had met the full medical board of eight members and that six refused to give up the right of election and two were in favour of transferring the right of future elections to the committee of management. Despite the intransigence of the majority of the medical board, the committee pursued the matter and drew up a series of rules to govern future appointments of medical staff. In essence, this was to ensure that the medical board's selection would go to the joint committee for confirmation. In a reply which once again quoted the 1815 Act, the medical board rejected these proposals.

On 25 March, in a move to show some authority, the joint committee sent a resolution to the medical board: 'That the long and faithful services rendered to the hospital by Assistant Surgeon, Dr Henry Stokes, will be duly recognised by the Medical Board when they proceed to elect a surgeon to the hospital in the room of William Pearson resigned.' Lambert Ormsby, who was a member of the joint committee, dissented from this resolution. William Pearson, a brilliant but difficult personality, had resigned in 1912 after being only a year on the staff of the Meath. He was later appointed surgeon to the Adelaide Hospital in 1915.[82]

The medical board did in fact elect Henry Stokes to be surgeon, on 11 April 1912, though Ormsby, Lane Joynt and Gogarty did not vote, wishing 'to take no part in the election' and, in communicating this fact to the joint committee, Ormsby read to them the following letter from Henry Stokes, which it was decided to record in the minutes (11 April 1912):

> Dear Sir Lambert Ormsby
> I have been given to understand through Dr Lennon that I will be elected a member of the Meath Staff next Thursday—I now write to tell you that I am sorry that any action of mine should be a source of discord between me and my future colleagues. I therefore wish you to understand that I see my voting against your election to the standing committee was not only ill-considered but hasty and of such a character as would naturally cause bad feeling. If I am elected I trust that the above subject will be dropped and that my colleagues and I will be able always to work together for the best interests of the Meath Hospital. I can now only express regret that the above incident ever occurred.
>
> Yours truly,
> Henry Stokes.

Ormsby's response is also recorded:

> In consideration of the explanation of the continued friction and dissension in the past, and the apology I have received, I have decided to withdraw my justifiable opposition on many grounds to Mr Henry Stokes; and I have prevailed on Mr Richard Vincent Slattery, Senior

Assistant Surgeon to the Richmond Hospital who had been strongly recommended to me as a most eligible and efficient Surgeon, not to be a candidate for the post of Surgeon at this election. In justice to myself and those who have acted with me, and also that my past and present action may be fully appreciated by the Conjoint Committee, I desire that a copy of Mr Stokes' letter and my explanation be inserted on our minutes and also forwarded to the Conjoint Committee for their information.

This extraordinary exposure by Sir Lambert Ormsby confirms William Taylor's statement that Ormsby had voted against Boxwell for the post of physician because he was a first cousin of Henry Stokes. He was apparently so resentful of Stokes that he voted for Edward Lennon's nephew, Charles Ross. Lane Joynt was probably obliged to vote with Ormsby, who was his senior surgical colleague. So this resulted in an unpleasant split of the medical board with a majority vote of three for Charles Ross, Ormsby, Lennon and Lane Joynt, and two for Boxwell, Sir John Moore and William Taylor.

After Henry Stokes' appointment, the tension between the members of staff and the conflict over the right of appointment between the joint committee and the medical board appear to have settled down. On 13 May 1912, after some consideration, the question of the joint committee's right to make appointments to the medical staff 'was postponed generally'.

During this period of tension, two appointments of interest were made by the medical board, though there was no specific confirmation of these in the minutes of the joint committee. Oliver St John Gogarty was elected as surgeon in February 1911. He filled the vacancy made by the retirement of Conway Dwyer, professor of surgery at the Royal College of Surgeons, who transferred to the Richmond Hospital. Gogarty was the first ear, nose and throat surgeon appointed to the Meath. His particular expertise was the treatment of sinusitis, and he had also recognised the importance of draining an unsuspected collection of pus in the maxillary antrum. Gogarty became best known for his brilliant literary ability, which completely overshadowed his medical career and his remarkable life has been fully described in the biographies by Ulick O'Connor and J B Lyons. He was a brilliant conversationalist, poet and wit, and author of *Tumbling in the Hay* and *As I Was Going Down Sackville Street*, and many other publications. He was also a senator in the first senate of the Irish Free State. He was notably the first Catholic to obtain a senior appointment at the Meath in the twentieth century.

Dr Robert Mathew ('Max') Bronte was appointed assistant pathologist in February 1912 and was promoted to pathologist in 1914. Born in Armagh and a member of the famous Bronte family of County Down, he was a pharmacist in Enniskillen before he took up medicine. He qualified from the College of Surgeons in 1906 and developed a special interest in forensic pathology. He was appointed 'crown analyst' for a time before he left Dublin for London in 1922. There he became famous as an expert in medico-legal work and gave evidence at many court cases, which included opposing the evidence of the famous Sir Bernard Spilsbury, the Home Office pathologist, in some notorious murder trials. In the biography of Spilsbury there are several uncomplimentary remarks:

> Bronte was the typical Irishman, as that abstraction is understood by the English—clever and quick-witted, voluble, combative, sociable, possessed of the gift of making friends and partisans . . .
>
> Celtic verbosity and cocksureness, the slipshod phrases and worse indiscretions that so often called down rebukes from the Bench, pampered rather than helped the defence, and did an injustice to Bronte's own genuine gifts.[83]

Max Bronte died of cardiac disease, at the early age of fifty-two, in 1932.

CHAPTER EIGHT

Infectious Diseases, 1894–1914

With the opening of the west wing ('epidemic wards') in the Meath in 1896, proper provision was finally made for the isolation of infectious fevers from the general medical and surgical cases in the main hospital. Typhoid was the commonest condition. Statistics for admissions to the west wing show that there were 79 cases of typhoid, with four deaths, in 1889–90. The number of typhoid cases fell over the following twelve years and in 1911–12 there were 25 cases and one death. Admissions of diphtheria cases continued throughout this period, with annual figures varying from 8 to a peak of 18 in 1910–11, and one or two deaths a year. Epidemics of measles occurred in 1889, 1902, 1904 and 1908, resulting in annual admissions of 40 to 60 cases. Scarlatina was common in 1901, 1902 and 1909, with up to 40 admissions in each of those years. The last recorded admissions of typhus were in 1889 and 1990. There was an epidemic of smallpox in Dublin in 1894–5. Such cases were admitted to the neighbouring Cork Street Fever Hospital.

Tuberculosis
Although the Meath provided separate accommodation for infectious diseases by the end of the nineteenth century, provision for the isolation and treatment of pulmonary tuberculosis was much delayed and limited. The public health authorities and voluntary agencies repeatedly brought pressure to bear on the Meath to assist in coping with the great number of cases of tuberculosis in the

community. In August 1901, Lord Powerscourt, chairman of the hospital board of superintendence, sent a letter to Sir John Moore, senior physician, requesting that special wards for phthisis be established. The medical board replied that 'it considered the allocation of separate wards for consumption in this hospital most undesirable and impracticable'.

In April 1905, Sir Charles Cameron, the medical officer of health, wrote to the joint committee referring to the great prevalence of tuberculosis in the city: 'My committee would earnestly suggest that in your hospital a small ward might be devoted to consumptives who would be retained until termination of their disease by recovery or death.' The joint committee replied as follows:

> Although our physicians are of the opinion that a general hospital is not suitable for the treatment of patients suffering from consumption, yet to meet the pressing need as far as possible, 33 cases of pulmonary tuberculosis were treated in our wards during the past year. Four of these cases died in hospital and several others refused to remain in hospital when they learned they were in an incurable stage. The remaining cases were either relieved or were sent to the Royal National Hospital for Consumption, Newcastle, or the Royal Hospitable for Incurables, and several of them were retained until they were admitted to Our Lady's Hospice for the Dying, Harold's Cross or the Rest for the Dying, Camden Row.

In October 1909, the Countess of Aberdeen wrote saying that a grant of £5,000 had been made by Robert J Collier of New York to the Women's National Health Association of Ireland to found a 'Tuberculosis Dispensary', and requesting a Meath hospital member to act on the relevant committee. The medical board nominated Sir John Moore, although their initial reaction was that the project was 'undesirable'. There was further correspondence between outside bodies and the hospital on this subject and in 1912 the medical board advised the joint committee 'to reply to the town Clerk to the effect with regret they do not see their way to make provision for sanatorium or dispensary treatment for tuberculous patients under the National Insurance Act 1911. More particularly any such arrangements should in the first instance be made with the Dublin County Council seeing that the hospital is the County Dublin Infirmary.'

Despite the continuing opposition of the medical board, the joint committee decided to send representatives to meet the insurance committee in the City Hall. At this meeting, enquiries were made as to the possibility of building on a 'derelict field to the west of the hospital'. The hospital representatives stated that the Meath would be willing to treat tuberculous patients if special accommodation was made available.

In April 1913, it was agreed that two wards of three beds each on the upper floor of the west wing would be reserved for TB cases. This was the first time an official decision had been made to provide special wards for active tuberculosis. Dublin County Council asked the hospital for a site for a TB dispensary and in May 1913 the joint committee decided to lease an area on the south-west corner of the hospital grounds, and to allow building there. Sir John Moore dissented from this decision. It was agreed that the lease should be for 200 years, from September 1913, at £40 per annum. The area was to be 'railed off' and a special entrance was to be made from Williams Place. The building of the Dublin County Council Central TB Dispensary was eventually completed in 1915. So it was that the Meath, though somewhat reluctantly, finally made a contribution to the enormous public health problem of pulmonary tuberculosis.

Changes at the Meath

In 1912 it was decided to appoint two resident qualified officers instead of the one resident surgeon, and that they would alternate as house physician and house surgeon every six months. The term 'clinical assistant' for annually renewable appointments was abandoned in favour of 'assistant physician' and 'assistant surgeon'. Thus the junior staffing became more distinctly divided into medical and surgical, whereas before duties had been more generally distributed.

The first X-rays at the Meath had been taken by surgeon Richard Lane Joynt in 1897. It was noted in the minutes of the joint committee on 30 October 1911 that 'X-ray installation is urgent'. The next reference to X-rays is on 14 October 1912: 'That Dr H W Mason be appointed radiologist at 20 guineas per annum and that Sister Nellie be given £5 per annum for keeping two X-ray rooms in order.' However, this arrangement was not a great success, as will be seen later.

The 'apothecary' had been a key post since the foundation of the hospital. As a resident in the hospital, and often senior in years and experience, it appears that he was usually given important administrative tasks, such as monitoring admissions of patients, as well as his normal duties of controlling supplies and dispensing medicines. He was also present or assisted at operations. In December 1912, it was decided to 'raise the salary of the apothecary (now resident pharmacist), Mr Tenison Lyons, from £50 to £70, in view of his long and faithful service and his voluntary service as Clerk of

Works in connection with building, plumbing and other structural alterations'.

Albert (Bert) Keating, who had acted for five years as assistant porter, was appointed 'gate porter' in 1913 in place of Thomas Bibby, who had retired after 27 years of service. Bert continued in office until 1963 and was the last to wear livery uniform—black frock coat with red collar and brass buttons. Bert lived for years in the porter's lodge and had to get up many times at night to open the gate. On one occasion he had to use a wheelbarrow to bring a patient from the gate to the casualty department!

Chapter Nine

War Years, 1914–19

Europe was precipitated into the Great War by the assassination of Archduke Francis Ferdinand at Sarajevo, and war was declared on 28 June 1914. Many Irish doctors volunteered for service with the allied forces and this seriously affected the staffing of the hospital. Geoffrey Fleming and E Bantry White, the two house surgeons, resigned to join the RAMC in August 1914, and Fleming was killed on active service six months later. Much of the work of the hospital during the war depended on medical students. Miss Nash, FRCSI, who remained for a year, was appointed resident medical officer with Miss G Murphy, senior resident student, to assist her, 'this being a national crisis and not to be a precedent'. In December 1915, Mr Cecil Robinson and Mr McKinnon were appointed senior and junior acting house surgeons for the duration of the war at the rate of £40 and £20 per annum, 'with rations to be supplied by the hospital'. Anaesthetist Dr Mason went to serve in the war, and anaesthetics were given in his absence by the house surgeons. William Taylor was nominated as surgeon to represent the hospital on the operation staff of the hospital at Dublin Castle and Edward Lennon was nominated as physician there. Henry Stokes was appointed to the Royal Military Hospital, Montpelier Hill. It was initially decided to reserve thirty beds in the Meath for sick or wounded soldiers, but only a few were occupied and the arrangement was not maintained after February 1915.

Accommodation of the Wounded

In 1916 accommodation began to be needed for wounded returned from France, and so beds were provided in the west wing for paralysed soldiers with incontinence, with attendants to be supplied by the military. In October, Henry Stokes was given leave by the hospital to serve with the RAMC at the front in France, where his famous brother Adrian Stokes, the bacteriologist, had been serving since the beginning of the war.

William Taylor, who had become president of the College of Surgeons in 1916, went to France to be chief surgeon in number 83 General Hospital at Boulogne in 1917. Cecil Robinson also went to serve in France and received a leg wound, in which the posterior tibial artery was severed. Amputation was being considered, when his teacher Lt Col Henry Stokes, unexpectedly turned up and saved his leg.

Fifty-two beds were placed at the disposal of the military authorities during the 1914–18 war. There were 626 sick or wounded soldiers of the expeditionary forces treated in the hospital and of these 4 died. The matron, Miss Bradburne, and Sister Margaret Lytle were awarded the Royal Red Cross for services to the sick and wounded.

In spite of the effects of the war and also the Easter Rising of 1916, the routine work of the hospital had to continue. In October 1917 the gynaecologist F W Kidd died and Paul Carton was appointed in his place. Kidd had been master of the Coombe Lying-In Hospital, but, despite his distinction, was not given the status of a full surgeon by the Meath, remaining instead as gynaecological assistant. Paul Carton was a Catholic, the second to be appointed to the medical staff after Gogarty. In 1923 he was promoted from assistant status to full senior level, after the death of Sir Lambert Ormsby.

Euphan Maxwell, the first ophthalmologist to the hospital, was appointed in June 1918. She had served with the RAMC in Malta during the war and also had a distinguished academic record. She was a sister of Constantia Maxwell the distinguished historian. She was also appointed to the Adelaide Hospital in 1922, and David Mitchell wrote of her that:

> she liked to be called Miss rather than Doctor, to emphasise her surgeoncy. She always dressed mannish as did so many women doctors of her time. The capacious side pockets of her tweed skirt were always in use, and with her well cut matching jacket she wore a collar and tie.[84]

Towards the end of the war in 1917, the provision of facilities for the treatment of venereal diseases was being considered. Dublin Corporation had made special arrangements with Steevens' Hospital and it was suggested that the

Meath should provide facilities for County Dublin patients. It was proposed that either the tuberculosis dispensary should be utilised or a special wing built for VD patients. However, no agreement was reached with the Meath and the County Council finally made arrangements with Sir Patrick Dun's Hospital.

In October 1918, during the influenza epidemic, the matron reported to the joint committee that there were 141 'bad' cases in the hospital. The housekeeper, four sisters, fourteen nurse-probationers and three servants were all laid up, and one died. There were only half the usual number of nurses available to run the hospital.

In April 1919 a vote of thanks was passed to Lt Col Henry Stokes for presenting two new sets of German surgical instruments, which had been obtained by Captain Adrian Stokes, DSO, during the war.

The 1916 Easter Rising

Scenes in the neighbourhood of the Meath Hospital in the week of the Easter Rising were described by Robert Collis in his autobiography. His father, W S Collis, a solicitor, was a member and former chairman of the joint committee, and Robert himself was a 16-year-old schoolboy at the time.

> My father, who was just back from the Italian front, had fixed the car up with red crosses painted on napkins and had put himself at the disposal of the Meath Hospital some days before, and I procured a couple of Red Cross armlets myself so as to assist him. During the afternoon I brought in a note to the Meath Hospital about some wounded and then went down to the hospital gate to see what was happening around. About two hundred yards away I saw a flag flying from the tower above Jacob's Biscuit Factory. It was waving in the breeze; a tricolour—green, white and orange. I stood silently gazing at it. A haze of brick dust hung in the air, caused by a stream of machine-gun bullets that were striking the tower below it. Now I walked down the back streets towards the factory. As I approached I found that its garrison was evacuating it and escaping into the narrow streets around the Coombe. Some were carrying tins of biscuits, others throwing out sacks of flour from the upper storeys. One of them came up to me with a revolver in his hand, and pointing to the Officers Training Corps badge that I had put in my button-hole to enable me to pass through the British lines and had forgotten to remove, remarked: 'You're just the sort of lad who gets shot if he doesn't look out, you know.' At that moment a bag of flour landed on a girl's head, and glad of a diversion I carried her into the hospital.[85]

Increased British military activity continued in the neighbourhood for some time after the rising and, in the minutes of the joint committee of 8 May 1916, it was noted: 'On Sunday the 7th, a very wet day, matron provided dinner for 68 officers and soldiers who were doing patrol duty round the hospital.'

The leaders of the Easter Rising were sentenced to death. Henry Stokes used to relate that, as consultant surgeon to the military authorities, he was asked to give a medical certificate that James Connolly, who was wounded with a compound fracture of the leg, was fit for execution. Henry Stokes was horrified, stating that he could never certify that a patient was fit to be shot, but they sat him in a chair and shot him anyway.

Chapter Ten

Towards Financial Crisis, 1920–30

Sir William Taylor was appointed regius professor of surgery, and surgeon to Sir Patrick Dun's Hospital, in 1922, and consequently resigned from the Meath. The medical board passed a resolution on 1 June congratulating him and deeply regretting 'the loss of one whose friendship has been as great an asset to the students and staff as his surgery has been an ornament to the records of the hospital'.

In September 1921 the matron, Miss Laura Bradburne, resigned after 32 years of service to the hospital. Miss Mary Wall was elected as her successor after a second election—in the first, the votes had been equal for her and Sister Margaret Lytle, who later became sister tutor. There was a serious clash later between the new matron and the secretary, Mr Dow, and both were summoned before the joint committee on 17 August 1922 and asked to co-operate.

While there were a few minor disagreements within the hospital during this period, outside of the hospital Ireland was plunged into dreadful civil war, following the signing of the Anglo-Irish Treaty in December 1921. At a meeting of the joint committee on 28 August 1922, a special resolution was passed:

> That the Joint Committee of the Meath Hospital and County Dublin Infirmary are, for the second time within a fortnight, under the painful necessity of expressing their deep regret at the death on active service of a

distinguished member of the Provisional Government of the Irish Free State, General Michael Collins, Commander-in-Chief of the National Army. The Joint Committee desire to convey to the relatives of General Collins and to his colleagues in Dáil Éireann their respectful condolence and heartfelt sympathy in their bereavement and on the loss which the country has incurred through so tragic an event.

At the same meeting, a letter was also read from the acting chairman of the Provisional Government acknowledging the resolution of sympathy passed by the joint committee on the death of Arthur Griffith.

There were resignations from the joint committee from the recorder, I L O'Shaughnessy, Dr P C Cowan and Sir Horace Plunkett. Plunkett, who had served on the joint committee since 1917, wrote a letter of resignation to the committee saying that he was obliged to leave Ireland 'now that my house is destroyed'. The founder of the co-operative movement, he had organised the co-operative creameries in Ireland. He had been unionist Member of Parliament for South Dublin from 1892 to 1918 and, although he later became converted to Home Rule, he opposed partition. Plunkett was a member of the new senate of the Irish Free State, but members of the senate were under threat of assassination by the anti-Treaty IRA. Plunkett's house, Kilteragh, in Foxrock, was burned down by the republicans in 1923.

Oliver Gogarty, who was also a senator, had been kidnapped by the IRA but made his famous escape by diving into the Liffey. However, his life remained in danger and in May 1923 he moved to London, where he continued his ear, nose and throat practice, returning to Dublin each week for meetings of the senate.

Hospitals' Sweepstakes

The annual report of the hospital for the year ending 31 March 1923 stated that:

> The year has been characterised by altered conditions necessitating drastic changes of policy, and by many disappointments. Bad trade and unemployment have affected the contributions from employers and employees alike, while heavy taxation and the unfortunate state of the Country has rendered it impossible for many of our accustomed annual subscribers to continue their support.

The hospital was now in grave financial difficulties. Costs were rising, with the development of radiology and pathology. Subscriptions and donations, which had reached £2,004 in the year 1920–1, had fallen to £1,170 in 1922–3, as a number of benefactors had left the country because of the political changes. The annual expenditure for the year ending 31 March 1923 was £35,162, and the income for that period was £12,161. A separate reserve account had a credit of £23,133.

The medical board supported the idea of running a sweepstake to raise funds. In March 1923 there was an offer by Mr R J Duggan to run a sweepstake in aid of the Meath Hospital, 'promoted from the Continent—not Ireland'. Sir John Moore proposed a resolution to the joint committee, seconded by Mr Good, not to proceed with a sweepstake 'until it is passed into law', but this was rejected. On 11 June 1923, the joint committee thanked Dr Edward Lennon for overcoming difficulties and bringing £10,000 to the hospital as a result of the sweepstake inaugurated by Mr Duggan, and the following June another £5,000 was received. These sweepstakes in aid of the Meath, and similar sweepstakes being run by other hospitals at the time, were illegal, but the government turned a blind eye to them in view of the serious financial state of the hospitals. They were forerunners of the legal Irish hospitals' sweepstakes which were set up by the Public Charitable Hospitals Act of 1930.[86]

Proposed Hospital Amalgamation

In January 1925 proposals for amalgamation were made by five Dublin hospitals—the Adelaide, the Meath, Mercer's, Sir Patrick Dun's and the Royal City of Dublin. The Rockefeller Foundation were interested in funding improvements in medical education in Dublin, and so two representatives from each hospital met to appoint a deputation to meet with the Foundation. The Meath nominated W S Collis, Wm Boydell and Henry Stokes to represent the hospital at the conference.

On 11 May 1925 the following resolution, proposed by Mr Collis and seconded by Dr Stokes, was carried by six votes to three:

> That in order to increase the facilities of treatment of disease, medical teaching, and research work, the Meath Hospital and County Dublin Infirmary are willing to agree with the Adelaide, Mercer's, Royal City of Dublin, and Sir Patrick Dun's Hospital for amalgamation of such five hospitals on such terms and conditions as hereafter may be arranged, provided that sufficient funds are made available for the erection of a fully equipped central hospital in Dublin with sufficient accommodation and endowment, and that necessary legal authority is obtained by each of

such hospitals to such amalgamation and the application of their endowment, income and property to such central hospital.

The majority opinion seemed to be in favour of working towards hospital amalgamation in 1925, but no further steps to that end were recorded. In February 1927 a letter was received by the joint committee from Dun's Hospital, which had been one of the prime movers, saying that amalgamation was not possible in the near future because of lack of funds.

In fact, during the 1920s, the new state was endeavouring to reorganise and improve the health services, with very limited resources, as the voluntary hospitals in Dublin were now in desperate difficulties.[87] This led to the legislation introducing the hospitals' sweepstakes in 1930 which saved the Irish hospitals from financial disaster.

There were still political tensions from the civil war and on 11 July 1927 the joint committee passed a resolution 'expressing horror at the callous murder of Mr Kevin O'Higgins, Minister for Justice. The Joint Committee would convey to the widow and child-daughters of the deceased minister their sincere sympathy.'

Staff Changes
Dr Thomas J D Lane, a Trinity graduate and a Catholic, was elected surgeon in place of Sir William Taylor by the medical board on 3 July 1922. There were five candidates including Dr Paul Carton, gynaecologist, with assistant status, and Dr Merrin, FRCSI, anaesthetist. Lane, who was to have an outstanding career in urological surgery, bringing international fame to the hospital, must have caught the eye of the senior medical staff at an early stage. Following his appointment as house surgeon to the hospital in 1920, he had been made assistant physician in July 1921. In that capacity he was partly engaged in X-ray work. The minutes of the medical board of 20 December 1921 record that 'an arrangement had been come to between Dr Lane and Dr Mason, whereby Dr Mason should carry out any X-ray treatment necessary at his private house, whilst Dr Lane undertook all photography at the hospital'.

Dr Henry Mason had been appointed radiologist (he was sometimes referred to as 'radiographer') in 1912. He also held appointments at Mercer's Hospital and Jervis Street Hospital, and was one of the pioneers in Dublin in raising money for the purchase of X-ray equipment.[88]

When Max Bronte resigned as pathologist in April 1922, T J D (Tom) Lane was appointed pathologist. So it was that, when he became visiting surgeon in July 1922, he also held the appointments of assistant physician, pathologist and quasi-radiologist. However, he resigned the post of assistant physician in August 1922 and Dr Cyril Murphy was unanimously elected in his place.

On 11 March 1924, the hospital secretary, Robert Dow, wrote to the secretary of the medical board:

> At a meeting of my Board held yesterday Dr Merrin raised the question of the three posts held by Dr Lane, i.e. Surgeon to the Hospital, Radiographer, and Pathologist. I have been directed by the chairman to ask you to bring the matter before your Board and let me have their views thereon. Dr Merrin considers that there should be an Assistant Surgeon appointed, and I should be glad to know their opinion on this matter also.

Dr C J Murphy, the honorary secretary, replied to the joint committee on 24 March 1924:

> I have been asked by the Medical Board to state that the members of the above Board are unanimous in considering that Dr Lane should hold position of Radiographer to the Hospital. With regard to his position as Pathologist, the present arrangement will not terminate until July next and meanwhile the Board will give the question its earnest consideration. The Board is unanimously of the opinion that it is quite unnecessary to appoint an Assistant Surgeon.

Tom Lane eventually resigned his post of pathologist in October 1924 and William Boxwell was elected in his place, and thus returned to the position he had held many years before.

The custom for all medical appointments at assistant status was for them to be renewed annually. This applied not only to assistant physicians and surgeons, but to specialist posts such as ophthalmologist, radiologist, gynaecologist and anaesthetist. Accordingly, the medical board decided that Dr Henry Mason should not be renewed in his post as radiologist in July 1924. Dr Mason strongly protested to the joint committee and the medical board, and complained of being treated unjustly. The medical board wrote to the joint committee that 'Dr Mason's appointment had become a mere sinecure and was costing the hospital £25 per annum'.

Though there is no formal note of appointment in the records, it is clear that from then on Lane was in charge of radiology. In October 1924 he was interviewed by the joint committee and he submitted a written statement on how the X-ray department should be run.

In December 1923 Sir Lambert Ormsby died, at the age of 74. He had been surgeon at the Meath for 51 years and had been a dominant figure. He was born in New Zealand of Irish parents and had received his medical education at both the College of Surgeons and Trinity College. Early in his career, Ormsby invented an ether inhaler which was used widely in anaesthesia for a number of years. He was elected president of the College of Surgeons in 1902, and was knighted in 1903. He had been apprenticed at the Meath to Sir George Porter. In the biographical sketch in his history of the Meath Hospital, 'ATL'[89] wrote that Ormsby's three watchwords were energy, perseverance and determination. Judging from his opinions in the minutes of the medical board, and his photographs, he certainly comes across as a very determined personality.

He had a special interest in orthopaedics, and he founded the National Orthopaedic and Children's Hospital in 1876, which later amalgamated with the Institution for Sick Children to form the present National Children's Hospital. He also had a great interest in introducing skilled hospital nursing. Together with Miss Ellinor Lyons, the lady superintendent of the Meath Hospital, he founded the Dublin Red Cross Nursing Sisters' Home and Training School for Nurses in 1885. This school was the first source of probationer nurses for the Meath, raising standards and attracting better educated women to nursing.

He published his first edition of the *Medical History of the Meath Hospital and County Dublin Infirmary* in 1888 and the second edition in 1892. The most valuable and interesting part of his book are the biographies of the physicians and surgeons of the hospital.

The affair of the chauffeur's boots, which took up the time of several meetings of the joint committee in 1922, shows a degree of megalomania on the part of the then 73-year-old senior surgeon. On 14 October 1922 it was reported to the joint committee that a patient's boots were missing. This would seem a relatively minor matter, but its importance became apparent when it was realised that the patient concerned, named Deegan, was chauffeur to Sir Lambert Ormsby. The sister-in-charge of the ward reported to the joint committee at its November meeting that the boots had been placed with the patient's clothes and taken away by his mother. However, Sir Lambert Ormsby did not accept the sister's report, having personally interviewed the wardsmaids. Sister Violet McLaughlin wrote a letter to the committee which was considered at the December meeting. She referred to the accusation by Sir Lambert that she was responsible for the disappearance of his chauffeur's boots and said that Sir Lambert had accused her, in front of the matron, of lying. Accordingly, she had paid 45 shillings to the chauffeur and was offering her resignation. The committee resolved not to accept her resignation and to reimburse her 45 shillings. Needless to say, Sir Lambert dissented from this decision!

Another break with the past going back to the Victorian era was the death of the apothecary, Tenison Lyons, in February 1923. He had given efficient and practical service to the hospital for thirty years. James Galashan was appointed in his place as dispenser, at a salary of 100 guineas a year.

Richard Lane Joynt, senior surgeon to the hospital, died on 8 April 1928. He had been a modest and retiring man, which was unusual for a member of the surgical profession. He was the first at the hospital to introduce X-rays, discovered by Röntgen in 1895, and through this work he suffered radiation damage to his hands. During the 1914–18 war he was appointed general inspector of orthopaedic factories in Great Britain and Ireland and adviser on the making of artificial limbs, and was awarded the OBE in recognition of his work for the disabled. From 1926 he also organised the tinfoil collection to raise funds for the Meath. Collection boxes were distributed throughout the country to collect silver paper wrappings from cigarette boxes, chocolates and other items. When these were collected they were melted down into ingots by Lane Joynt in the basement of his house at 83 Harcourt Street. A finished ingot was worth £13.

Lane Joynt was one of the first to own a motor car in Dublin. On 27 November 1902, he was summonsed for driving on the Stillorgan Road at a dangerous rate. In cross-examination, the police sergeant said that the doctor was driving at a rate beyond six miles an hour. Lane Joynt estimated that his speed was 12 miles an hour, and the magistrate dismissed the case! Lane Joynt was a literary scholar and was able to quote Shakespeare at will. He was also an enthusiastic yachtsman and was immensely popular with his students.

At the medical board meeting on 16 March 1928, it was proposed by William Boxwell, seconded by Sir John Moore and passed unanimously that Cyril Murphy should be elected on to the senior staff in succession to Lane Joynt. Cyril Murphy had been assistant physician for six years. It seems unusual for a physician to be appointed in place of a surgeon. This meant that now there were four physicians to four surgeons on the medical board, whereas there had always been a majority of surgeons in the past. The physicians were Sir John Moore, Edward Lennon, William Boxwell (who was also pathologist) and Cyril Murphy. The surgeons were Oliver St J Gogarty, Henry Stokes, Tom Lane and Paul Carton. Gogarty was an ear, nose and throat specialist and Carton was a gynaecologist. Lane would increasingly specialise in urological surgery, which would leave Stokes to carry most of the load of general surgery. However, the pressure on the surgical side was relieved in August 1928 by the appointment of W Collis Somerville-Large, an orthopaedic specialist, and Sydney V Furlong, an ear, nose and throat specialist, as assistant surgeons. Cecil Robinson and Michael Cuffe were appointed assistant physicians at the same medical board meeting. Cecil Robinson became honorary secretary of the medical board a year later, in 1929, and continued faithfully in that office until 1964. Dr Cuffe's appointment, however, was not renewed in 1929.

In September 1927 Adrian Stokes, brother of Henry Stokes and grandson of William Stokes, died in West Africa of yellow fever, which he had been investigating. He had been a student of the hospital and subsequently had had a brilliant career. A decorated scroll commemorating Adrian Stokes' work was sent to the hospital by the Rockefeller Foundation.

In January 1930, W J E Jessop, professor of physiology in the College of Surgeons, was appointed as biochemist to the hospital, thus beginning a long association with the hospital. His honorarium was agreed in 1931 to be 50 guineas a year. Dr Silvia Deane Oliver, who had been part-time anaesthetist since 1929, was also appointed full-time at 50 guineas a year. She became a well-known personality in the hospital, a dignified and efficient lady who used to come to the hospital in a pony and trap right up to the late 1940s. She was put in charge of the medical records in 1935.

By now the medical work involved less infectious fevers and, in 1930, it was proposed that wards E and F in the west wing be used for non-infectious medical cases. This was stoutly opposed by Sir John Moore, who pointed out that the annual grant received specifically for the care of fevers since 1827, namely £600 per annum, was now absorbed into the general income of the hospital. But it was to no avail, and the changing pattern of disease meant that the west wing would never again be reserved purely for infectious diseases.

Chapter Eleven

The Dormant Thirties, 1930–9

In September 1930 the joint committee agreed to take part in the Irish hospitals' sweepstakes, with Sir John Moore dissenting. Amongst other hospitals in Dublin which also agreed to take part were Dun's, Jervis St, Holles St, Harcourt St Children's, Teac Ultain and the Dental Hospital. 'A committee of reference' was set up by the government to receive applications and to decide on allocations of money to the hospitals from the proceeds of the sweepstakes. The sweepstakes were an enormous success and there was a rush by each hospital to get as much money as possible. A shopping list was prepared by the Meath, which included:

Upgrading the out-patient dispensary	£15,000
New boiler house and chimney	£8,000
Laundry extension south of the west wing	£14,000
Building for medical officers	£4,000
Building for servants	£7,500
X-ray department	£6,500
New operating theatre and steriliser	£3,200

Grants were also claimed for furniture, repayments of loans, etc. The total grant which the Meath applied for initially was £188,319 15s 7d, and £150,406 3s 10d was awarded [report of Committee of Reference, 1931]. This was a tremendous windfall to the hospital and led to an epidemic of repairs and building work. The dispensary building was improved and Gogarty seized the opportunity to plan his own clinic at the west end of the building with an escape door. When tiring of the patients which were crowded in the waiting area, he would slip unseen out of this door and ride off on his bicycle. As he personally knew the hospital architect, Mr Dunlop, whose office was next door to his own house in Ely Place, he was able to arrange to have special attention, such as having the pipes in his clinic chromium plated!

The enormous chimney was erected at the back of the hospital, despite complaints by the neighbours. This was for coal firing and was probably not necessary for subsequent oil burning. Residents' and maids' quarters were constructed along the west boundary, and general improvements were made to the wards, X-ray department and operating theatre. No building on the south side of the west wing was started, but the project remained on the agenda for some time.

In the next report by the committee of reference of 1932, the total list of claims by the Meath amounted to £265,410 8s 2d and the total amount awarded was £262,496 16s 5d. A plan for extension of the west wing for 61 new beds was submitted, but this was deferred because it exceeded the official overall hospital bed allowance.

Appointments and Resignations
In October 1932, Dr W R F (Bob) Collis was appointed assistant physician. He had graduated at Cambridge and had been trained in paediatrics in London and at John Hopkins in Baltimore. He had also done research work on rheumatic fever and erythema nodosum. There was only the small Smyly children's ward in the Meath and there does not appear to have been a great need for a paediatrician. No doubt the fact that his father was chairman of the joint committee would have had some influence, but he was already well known as a multi-talented doctor who would be able to stimulate research. His other hospital appointments in Dublin were purely paediatric—at the Children's Hospital in Harcourt Street and the Rotunda's neo-natal department.[90] In his second autobiography, Bob Collis described his colleagues at the Meath in 1932, as he remembered them.

> It had the most unusual staff. Sir John Moore was the oldest member, somewhere in his eighties. Next came Dr Lennon. Nobody knew his age but it was said that Sir John could remember things clearly which had happened forty or fifty years before. Then there was Dr Boxwell who was a physician and pathologist, and was great friends with the matron who

was up-stage-and-country English. On the general surgical side was the last of the Stokes, one of the most famous medical families, with a disease called by their name. Henry Stokes, the present family representative, was a gentleman of the old school.

He was too kind to charge the poor any fees and too grand to charge his friends. In consequence he was exceedingly poor. I once heard an old woman from the Dublin slums describing him to a friend, 'Ah, Mr Henry Stokes,' she said, 'there's no rhyme or reason in him, he's like the love of God.' Gogarty, poet, literateur, and talker known all over the world, was ear, nose and throat surgeon, with a great reputation in every direction. He was said to be a surgeon of ability, though his method of chopping off tonsils rapidly without an anaesthetic frightened me. Lane, though a physician, was a urogenital surgeon of considerable renown. There were also a number of younger men.[91]

His description of the staff, though written many years after he had left Dublin, seems very accurate. There are those alive still who remember assisting Gogarty, and were struck by his alarming lack of feeling for his patients during operative procedures such as tonsillectomy without anaesthesia.

Collis described the 'ancient board room with its portraits of famous old doctors and illustrations of early operations, where the staff foregathered at eleven o'clock each day'. This custom of all the medical staff meeting for coffee at eleven continued up to recent times. Apart from social contact, it provided an opportunity to consult with one's colleagues, especially the radiologist or laboratory specialists, on particular cases. Everyone disappeared back to their work within half an hour, except for Dr Cecil Robinson, honorary secretary, who for many years would hold the fort there during the morning, ready to act as father confessor and collecting student fees, keeping the accounts of the medical board and writing up the minutes, and planning the next annual dinner.

As the sweepstake funds were stimulating thoughts of development, Bob Collis put forward plans in 1933 to develop the west wing as a chest unit which would treat tuberculosis, bronchiectasis, empyema, pleurisy, etc. This seemingly progressive idea was not received well, with Edward Lennon opposing the idea 'unless the special unit was in a separate building'. Bob Collis continued at the Meath on an annual appointment as research assistant until 1936, when he left to devote his time to the Rotunda and the Children's Hospital.

Not all new ideas had such an unfavourable reception, and in April 1933 a consulting psychiatrist was appointed for the first time. This was Dr Richard Leeper, who had been medical superintendent of St Patrick's Hospital since 1899. He was given the old-fashioned title of 'consulting alienist'. He was succeeded by Dr Robert Taylor, and later by Dr Norman Moore in 1946, the latter holding a busy weekly psychiatric clinic at the Meath.

Some other useful ideas were also adopted. Dr Jack O'Leary was put in charge of the casualty department in 1936. Jack O'Leary was a general practitioner with a large practice living at the corner of Heytesbury Street and the South Circular Road. He had been a student and a house officer at the Meath and knew the hospital and the surrounding neighbourhood intimately. He was full of shrewd commonsense and pragmatism, and was an ideal person to advise and teach the inexperienced house officers and students in the extern department. His special relationship with the Meath was revealed by his method of referral of patients for admission to the hospital. Medical cases would be sent with a note addressed to the hall porter: 'Dear Bert, put this patient under Brendan O'Brien, Yours Jack', and Bert Keating would get the patient admitted to the appropriate ward. Jack O'Leary will be remembered by many students with gratitude, and also by many of the poor to whom he was very generous.

Tom Lane continued to develop the X-ray department and had received grants towards the purchase of a portable X-ray machine and other improvements. As he was also an active surgeon, becoming more and more involved with urological surgery, he was under considerable pressure of work. In February 1933 the medical board appointed Dr Cyril Murphy to be medical radiologist and Mr Tom Lane to be surgical radiologist. In this way Murphy reported on the chest films and Lane on the other X-rays. In 1934 the resident secretary, Robert Dow, was moved out of his quarters on the ground floor of the Barber wing to make way for the expanded X-ray department. Lane was very unhappy at the high mortality rate associated with the open methods of prostatectomy then being practised. In 1938 he visited the Mayo Clinic and found that they had a much lower mortality rate, due to the transurethral prostatectomy method which had been developed there by Dr Gershon Thompson. When he returned to Dublin he began the punch resection of the prostate, greatly reducing the mortality. He had also been advised on the imperative need for having specially trained urological nurses. The joint committee noted that 'a larger number of genito-urinary cases are treated in the Meath, we believe, than any other hospital of its size in Ireland', and it was agreed that special nurses should be appointed for these cases.

Dr Edward Lennon became ill in 1937 and was unable to attend the hospital. A resolution from Alderman William O'Connor of the joint committee was received by the medical board at its meeting on 17 November 1937 'that the Medical Board be asked to advise as to the desirability of appointing a doctor temporarily to carry on the skin dispensary during the regrettable absence of Dr Lennon owing to illness'. The medical board replied that the 'question of making arrangements for the continuance of the skin dispensary is under consideration and will be dealt with at an early date'.

Dr Herbert Mackey, a general practitioner in Dun Laoghaire who had a special interest in skin diseases and had even written a book on the subject, applied to the medical board in December for the post rendered vacant by the illness of Dr Lennon. However, Dr Lennon had made arrangements for Dr Augusta Young, an assistant in the dermatological department of the Adelaide Hospital since 1929, to carry on the work during his absence.

Sir John Moore had resigned in 1933 at the age of 88, after 58 years of service at the Meath. He had always had a special interest in fevers and was a direct link with the era of William Stokes—he had been Stokes' last house physician—whom he succeeded in 1875. He was closely involved with the formation of the Royal Academy of Medicine, which resulted from the amalgamation of various medical bodies in the city in 1882. He was president of the College of Physicians in 1898 and its representative on the General Medical Council from 1903 to 1933. He received honorary degrees from Trinity College, Oxford and the National University. In October 1934, an exceptional resolution was sent to Sir John from the Dublin City Council 'that the City Manager convey to Sir John Moore on our behalf an expression of the City Council's deep appreciation of his services given over a period of 59 years in the interest of the health of the citizens of Dublin'.

Apart from his many great distinctions in medicine, he had a great interest in astronomy and was one of the foremost meteorologists in Dublin. He had instruments for measuring the temperature, barometric pressure and the rainfall on the open balcony at the end of ward 12 and would include a visit there on his ward rounds to make his readings. Even after his retirement he continued to come to the hospital daily to take measurements of the rainfall, etc. He died in 1937 and Dr Robert Steen was appointed physician in his place. Steen had a special interest in children's diseases and he was to resign from the Meath in 1948 to restrict himself to paediatrics at the Children's Hospital.

In April 1937 Richard Leeper, consulting alienist to the Meath and medical superintendent of St Patrick's Hospital, wrote a furious letter to Henry Stokes, which was considered at the medical board on 18 May and marked 'read'. Apparently it had been suggested, presumably in jest, that the distinguished surgeon to the hospital from 1849 to 1895, Sir George Hornidge Porter, was a son of the hall porter. This must have struck a sensitive nerve in Dr Leeper who, as can be seen in his letter, was a relative of the Porter family.

> I was greatly shocked to know that it was stated that Sir George H Porter was the son of a hall porter, and that such was figured in a coat of arms of the family of Porter. I am amazed that such a figurative evidence of gross ignorance should be stated in the Meath Hospital—of all earthly places.
>
> My father was a House Surgeon of the Meath Hospital, and I hold his autograph letters stating this fact. He was a fellow student with George Porter, son of William Henry Porter, Professor of Surgery in the College of Surgeons. He entertained Dupuytren in his house in Kildare Street, and, when he was a surgeon to the Meath Hospital, lived at his country residence, Tibradden House, now lived in by Col Guinness, from which he drove down to the Meath every morning with his coach and horses. Think of this as a comparison with the surgeons to the Meath today. In those days, the profession was very different to the comparative squalor of today. I hope you will erase this insult to the Porter family from the Meath traditions and set it right, forever.
>
> . . . I have a portrait of my aunt—Sir George Porter's sister and shall allow you to copy it, if you really want to discredit this damnably libellous statement, that Sir George Porter descended from anyone but from his own father—William Henry Porter, Professor of Surgery in the Royal College of Surgeons in Ireland.

In 1938 Dr Paul Carton, gynaecologist, died. Dr James Quin was elected in his place and eight beds were designated for his patients.

In November 1938 Oliver Gogarty presented an elegant bronze bust of himself and a table which had belonged to his father to the medical board. The bust and the table have been in the old boardroom of the hospital ever since. These gifts to his colleagues seem to have been in preparation for his resignation from surgery and the Meath in order to devote himself entirely to his literary career. A year had elapsed since he had lost the famous libel action taken against him for malicious comments in his *As I Was Going Down Sackville Street*. He resigned from the Meath in a letter written from London on 10th February, 1939 to the medical board:

> *11 Mansfield Street, WI*
>
> Dear Dr Robinson
>
> As it might have helped in a lawsuit to be attached even nominally to the Meath Hospital, I postponed my formal resignation. I find that the case may be as late even as two months from the present date. Therefore it is hardly fair further to postpone my resignation which I hereby send.

It is not necessary to say with what regret I sever a connection of 29 years with a hospital so eminent for its long tradition and high decorum in medicine.

To my colleagues, old and young, whose friendship never failed me, I extend a hand in farewell.

Yours sincerely,
Oliver St J Gogarty

In this letter he is referring to a libel lawsuit which he was taking against the publishers and printers of Patrick Kavanagh's *The Green Fool*. He was eventually awarded £100 damages and, some months later, he left Ireland for the USA, where he continued his brilliant literary output. Whenever he returned to Ireland he would visit the Meath Hospital to see Tom Lane. He died in New York in September 1957 and was buried in his beloved Connemara.[92] Tom Lane proposed that Sydney Furlong, who was assistant ear, nose and throat surgeon, should be elected in Gogarty's place, and this was passed unanimously.

CHAPTER TWELVE

The Emergency Years and Tuberculosis, 1939–48

With the outbreak of the Second World War, despite Ireland's neutrality, plans were made in 1939 to deal with the possibility of air raids. Eight Dublin hospitals were designated as casualty clearing stations, including the Meath, which had a commitment to provide 50 beds if required. Casualties were to be evacuated, after initial treatment, to a '400 bed evacuation hospital outside Dublin'. The staff were trained in air raid precautions (ARP), but, fortunately, Dublin remained virtually unaffected by the 'Emergency', except for some shortages of supplies and fuel.

In the 1940s, Ireland was in the grip of a tuberculosis epidemic. There were not enough beds in the tuberculosis sanatoria, and many cases, some of them open and infectious, were being admitted to general hospitals. A significant number of nurses recruited from rural areas were non-immune and contracted pulmonary tuberculosis during their probationary years. The arrival of Brendan O'Brien influenced the measures taken by the Meath to meet this problem. The joint committee agreed to construct a wooden hut at the rear of the west wing for the isolation of TB cases. The Hospitals' Commission stated that 'they were not prepared to sponsor development of this nature' and so the ladies' committee agreed to raise the necessary money, and they decided it should be called St Mary's. In 1944 the building was constructed, linked to the back of the west wing, and in the same year it was agreed that the two top wards of the west wing should also be reserved for tuberculosis cases.

The joint committee were looking for more accommodation for the Meath, either for administration, or residences, or an auxiliary hospital. They were unhappy, from the point of view of liability, about the number of patients from outside Dublin attending the genito-urinary department who were residing in 'digs' in the neighbourhood. Enquiry was made about the possibility of acquiring St Peter's School, an old building diagonally opposite the hospital gate which belonged to the Church of Ireland. However, the purchase did not proceed because the hospital was legally advised that, under the charter, it was 'ultra vires' to purchase property outside the grounds.

The matron, Miss Winifred Gage, who had inspected the school building with members of the joint committee, resigned in March 1946. She was a dignified lady who had been released for war service for a period in 1943–4, during which time Miss Ann Magee had been appointed matron in her absence. Miss Magee was now appointed matron once again, at a salary of £225 per annum.

Edward Lennon, physician, died in September 1940. He had served the hospital, including his time as assistant physician, for over 50 years. He was known for his financial acumen. It was Lennon who had been instrumental in obtaining money for the building of the west wing in the 1880s and had saved the hospital from financial disaster by organising sweepstakes in the 1920s.

On 26 September 1940, William Boxwell proposed and Cyril Murphy seconded the proposal that W J E Jessop, biochemist to the hospital, be elected physician in place of Lennon. This was unusual in that Jessop was not a clinician. He was a professor of physiology at the College of Surgeons and the argument used in favour of his appointment was that he would teach the application of physiology and biochemistry to clinical problems at the bedside. This he did for many years, at a weekly clinic, using patients under the care of one of his clinical colleagues.

Professor Jessop was appointed to the chair of Social Medicine (later Community Health) in Trinity College in 1952. He became dean of the Medical School there and was largely responsible for its revival in the 1960s.

In 1942 Mr J I Fitzpatrick was appointed manager-secretary to the hospital. He revised various accounting procedures and, in the annual report of the hospital for the year ending 31 December 1943, gave an impressive report on 'Steps taken to reduce expenditure and increase income and generally to improve the efficiency of the working of the hospital'. This report showed a remarkable number of reforms in all aspects of the hospital where money was involved, resulting in reducing the annual expenditure in 1942 from £34,744 to £29,561 in 1943. Income was increased in 1943 to £21,176, from £16,971 in 1942.

In January 1943, Dr Herbert Mackey asked permission of the medical board to attend Dr Young's skin dispensary as her clinical assistant. The board replied that he could attend for six months but that there was no such post as clinical assistant available. This agreement was renewed at subsequent meetings of the medical board up to 1948. Dr Mackey also acted as locum at the skin dispensary during Dr Young's absence.

William Boxwell died in May 1943, leaving vacancies for a physician and a pathologist, as he had covered both posts; he had also been professor of pathology in the College of Surgeons. He had been elected president of the College of Physicians in 1937, and at the same time Henry Stokes, his first cousin, was president of the College of Surgeons. Both were grandsons of the famous William Stokes.

The medical board elected Dr Brendan O'Brien as physician in place of Boxwell. Prior to taking up medicine he had been a coffee planter in Kenya and later was district commissioner in Uganda. He was involved with the colonial service when he contracted malaria and tuberculosis and, after some time spent in a sanatorium in Switzerland, he developed an interest in medicine and entered the medical school in Trinity in 1930, graduating in 1935. O'Brien had become assistant physician at Baggot Street Hospital and also assistant to V M Synge, professor of medicine at Trinity College. He was a great-grandson of the nationalist William Smith O'Brien, who was at first sentenced to death in 1848 but later transported to Tasmania. He was son of Dermod O'Brien, president of the Royal Hibernian Academy, and his mother, Mabel Emmeline, was daughter of Sir Philip Crampton Smyly.[93]

Being grandson of Philip Crampton Smyly was too remote a connection, if appreciated at all at the time, to have any influence in the appointment. In fact, apart from his obvious good standing in the profession and teaching ability, he was particularly well suited for the Meath by his experience in the treatment of pulmonary tuberculosis, which was a major problem at the time and in which he became actively engaged at the Meath.

To cover the duties of pathologist, the medical board appointed Professor John McGrath, who agreed to provide a service from his laboratory at St Vincent's. This arrangement proved too difficult for him because of his many other commitments and only lasted a year, after which Professor Jessop agreed to do the biochemistry and clinical pathology and Professor O'Meara of Trinity College was appointed honorary bacteriologist.

Planning the Genito-Urinary Unit, 1946

In October 1943, Mr Tom Lane gave the inaugural address [94] at the opening of the 1943–4 clinical session. Lane, who had then been surgeon to the hospital for 21 years, reviewed the remarkable advances in medicine and surgery since 1920, especially in the treatment of diabetes, modern anaesthesia and the beginning of specialisation in surgery in the Dublin hospitals. Though he did not specifically refer to it, Lane himself had already developed a reputation for specialisation in urological surgery, despite limited resources.

Lane went on to mention the hospitals' sweepstakes, which had produced an astonishing amount of money for hospital development but had been brought to an abrupt end by the war, though a large sum had accumulated in the Hospitals' Trust Fund. 'This fund has now ceased to increase, and with the Hospitals' Commission plan for four large general hospitals for Dublin frozen for years to come, the moment has come for a complete review.'

As an example of the expansion in technical work which was bound to continue, he quoted figures from the X-ray department of the Meath. In 1929, some 1,500 patients were X-rayed, at a cost to the hospital of £183. In 1939, about 5,300 patients were X-rayed, at a cost of £947. To deal with the shortage of beds, he made a strong plea for the establishment of a lightly staffed auxiliary hospital to which patients could be discharged a few days after surgery. This was an unusual idea at the time but could have application to this day. He also emphasised the need for development of laboratory services in pathology, biochemistry and bacteriology.

Finally he complained bitterly of the rates of pay for nurses:

> A year's arduous work enables a senior sister to have her annual holiday and very little else. There is practically speaking no pension scheme, and yet as we all know nurses are the backbone of the hospital.
>
> ... If we have to choose now, as I think we have, between new buildings on the one hand and proper staffing on the other, I would unhesitatingly vote for proper staffing.

Tom Lane had an increasing number of patients being admitted under his care, most of them requiring prostate surgery. However, he did not have enough beds and so many of his cases overflowed into other surgical wards. In 1944 he gave up his X-ray work and Dr Harold Pringle was appointed the first specialised radiologist to the hospital.

Henry Stokes was now the only senior general surgeon and with his increasing years he needed assistance. In February 1945 Henry Stokes proposed, and Tom Lane seconded the proposal, that R F J (Jack) Henry, of Baggot Street

Hospital, and Douglas Montgomery, of Dun's Hospital, be appointed reserve surgeons, this to be renewed on an annual basis. This was an unusual title and presumably reflected the need for backup for general surgery. It would appear that these appointments may not have been enough for emergencies and so Brandon Stephens, assistant surgeon at Baggot Street Hospital, agreed to assist Henry Stokes. The very night of his interview with Stokes, he was called to the Meath to perform an emergency appendix operation!

The war ended in 1945 and, as the general situation relaxed, conditions improved for planning development. Tom Lane's reputation in prostatic surgery had now spread, but the great volume of his genito-urinary surgical work was being done in very cramped conditions. The medical board resolved, in December 1946, that they were 'in favour of establishing a separate GU (genito-urinary) unit on condition that the unit be sufficiently large to house all GU cases and that all such cases be confined to the unit'. This heralded the biggest change in the work and structure of the hospital since its foundation two hundred years before.

The plan that emerged was to construct an L-shaped building along the Heytesbury Street and Arnott Street boundaries. This was to provide a new nurses' home and also, originally, a section for hospital administration. This would free the old nurses' home, which was already linked with the main building, for conversion into a GU department. This was a substantial building, of the same style as the Barber–Bury wing, linked by a glass covered bridge with the main building at first floor level.

On 15 November 1948 the joint committee received a letter from the Department of Health agreeing to this plan and cancelling an alternative plan for a 30-bed auxiliary unit which had been proposed by Tom Lane to relieve pressure on surgical beds. Some days later, Tom Lane sent a generous and far-sighted letter to the medical board:

> I would like to take this opportunity of thanking my colleagues on the Medical Board for their help and encouragement through so many long years and would like to single out for very special mention Henry Stokes for his generosity, tolerance, broadness of outlook, and above all for his kindness. It is impossible for me when thanking the living not to have very much in mind the dead—Lane Joynt, a very particular friend, Lennon, who was so very kind, and Boxwell, from whom, in spite of petty estrangements, I received so much solid help as reference to the minutes of the Medical Board over the last dozen years will easily prove.
>
> ... It is obvious that the next few years will throw a very considerable extra strain on me personally. In addition to organising and planning the

new department which will be a small specialised hospital in itself, I will have to continue to work at high pressure to maintain and deal with the flow of work done under such unfavourable and cramped conditions as exist in the present GU department. It is essential to encourage, maintain, and, if possible, to increase the flow of work in order to give the new department a flying start. The first important step was the appointment of Dr Mayne to the department.

I am constrained and compelled to ask my friends of the Medical Board to deal promptly with the out-of-date and unreal situation with regard to general surgical emergencies for which I am supposed to be responsible every alternate week. The GU department is geared up to do a highly specialised type of work for which it has managed so far to refuse help to none in urgent need. The equipment of the department has cost thousands of pounds. As you are aware, it has a team of specially trained nurses provided at considerable expense and, in addition, has the part-time service of Dr O'Leary and the whole time service of Dr Cussen for which I myself pay—services available to public and private patients without any discrimination. It seems to me cruelly unfair that I and those who work with me should be expected to make any contribution in this last round of our battle towards the problem of general surgical emergencies. I do not deny for a second that I was appointed surgeon to the hospital as a general surgeon. I maintain that without your help I would not be what I am now. I feel sure I can rely on your support in this, these last difficult days.

It is pertinent to point out that if the new GU unit has to be planned, the Medical Board, for its part, will have to give the most careful consideration to the situation which arises when the GU department is evacuated to its new home. In any general review of the position of the hospital from a surgical point of view I believe it is most important (both from a national and urological standpoint) that no effort should be spared to secure the preservation and continuation of Mr Stokes' pioneer work on the thyroid side by side with the development and encouragement of abdominal surgery. The provision of a chest unit should be seriously considered and I personally would be very glad to see Mr Henry working with our physicians.

To enable this to be done on a proper scale, it may well be wise to reconsider federalisation with Dr Steevens' Hospital and very possibly too, a close and professional and educational affiliation with the neurological department of the Richmond. It seems to me that if these things could be achieved, the surgical staff at the Meath together with the surgical staffs of Steevens' and the neurological department of the Richmond will offer a wide and well-organised surgical service to students and patients alike.

The present Minister for Health has thrown down a challenge to all sections of the medical profession in this country demanding that it provide the best possible medical service to our countrymen and stated that only by its so doing will state control be avoided ... I feel certain that this great little hospital with its high sense of service and pliability of outlook will play a leading part in taking up the Minister's challenge.

Dr Brian Mayne, mentioned in this letter, had served in the RAMC during the war and had been taken prisoner by the Japanese, enduring terrible suffering for three and a half years. He returned to Ireland and took his membership of the College of Physicians. He was appointed clinical assistant to the physicians in June 1947 and clinical assistant to the GU department in July 1948. He rapidly became known for his remarkably sharp brain and for his efficiency, and he made a great contribution in the medical assessment and care of elderly patients undergoing surgery in the GU unit. Dr Mayne was elected physician to the hospital in October 1948 in place of Dr Robert Steen, who had resigned to confine himself to paediatrics at the Children's Hospital, Harcourt Street. Mayne was a popular teacher and taught applied physiology in the College of Surgeons for many years. He developed a very large private practice and had many distinguished people under his care. He was a grand-nephew of Robert St J Mayne, the physician to the Meath who died of smallpox in 1871.

Towards the end of 1948 all seemed peaceful. Plans had been agreed for the development of the GU unit, including its own separate records department, a new matron had settled in and the building of the new nurses' home was soon to commence. It was agreed that the Lord Mayor be asked to take the chair at the Triennial General Meeting on 4 April 1949, when the election of the new joint committee would take place. The manager-secretary would arrange the proposing and seconding of the necessary motions. These routine matters marked a calm period before the unexpected storm that was about to hit the hospital.

Chapter Thirteen

The Advance of the Knights, 1949–50

The Act of Parliament of 1815, under which the Meath Hospital and County Dublin Infirmary was constituted, stated that any donor of a sum not less than twenty guineas would become a governor for life, and any person who subscribed two guineas would become a governor until the next 1 January. It also stated that an annual meeting of the governors would be held every first Monday in April to elect the twenty-one members of the standing committee to manage the hospital. However, since the Local Government Act (Ireland) of 1898, which specified that a county infirmary receiving a grant from a County Council should be managed by a committee elected triennially, the annual general meetings took place only to receive a report and accounts for the year, except for once every three years when the election of the joint committee was held.

The joint committee had been elected every three years by ballot from 1898 to 1929 and the custom had developed of re-electing the committee en bloc. New members of the committee that had been co-opted in place of those that had died or resigned became confirmed as elected at the triennial election. At the joint committee meeting on 17 February 1947, it was decided that, as the last election had been on 1 April 1946, the next would be held in 1949. Accordingly, the meeting took place on 4 April 1949. There have been many accounts of what occurred at this meeting from different individuals, witnesses in court and newspaper accounts. The essential points are as follows.

The triennial meeting was advertised in the daily newspapers in accordance with the Act of 1815 and the agenda was given as 'election of Joint Committee'. Certain governors who were noted down to speak received by post a copy of the agenda which had as item three 're-election of outgoing Joint Committee'. Tea was at 4 o'clock in the front hall. There was an attendance sheet at the door and most, but not all, of those present signed. It was noticed that there were about thirty strange men never previously at the meetings of the hospital present at the tea. They all wore dark clothes and some thought that they were connected with a funeral. Major Kirkwood, chairman of the joint committee, said in evidence at the court case later that he thought that they had come to pay tribute to the late Lord Meath, about whom a motion was standing on the agenda. After tea, the attendance adjourned to the nurses' home for the meeting. There was no check on the identity or credentials of those present.

As was the custom at triennial meetings, a distinguished person, not necessarily a governor, was proposed to take the chair. The Lord Mayor of Dublin, Alderman John Breen, who was not a governor, took the chair at this meeting. Before the election of the outgoing committee, which was to be proposed by Major Kirkwood, was called, a stranger stood up and proposed the name of another stranger for election to the committee. There followed a spate of nominations of strange names which were rapidly proposed and seconded. There were a number of protests, with several people being on their feet at the same time. Henry Stokes and Cecil Robinson objected to the rushed proceedings and to being asked to vote for people they knew nothing about. Mrs Mulvey asked the Lord Mayor to allow the meeting to stand adjourned for legal advice, as she believed there was something irregular about what was happening.

The Lord Mayor asked everyone who was not a governor to leave the room and he eventually ruled that the election should proceed. Pieces torn from a notebook were used as ballot papers and the Lord Mayor ruled that they should not be signed and that the ballot should be secret. Three scrutineers were appointed and 53 votes were counted, though some said that there were only 46 in the room. The Lord Mayor had to leave the meeting early and the result of the ballot was read out by Major Kirkwood. Thirteen strangers were elected to the new joint committee, which contained only one representative of the medical board, namely T J D Lane.

The nineteen members elected to the new joint committee were as follows: Reverend Father William Kealy, W T Cosgrave, Senator Michael Colgan, Dr Angela Russell, Michael Devlin, Desmond McAteer, Alphonsus Lyons, Senator S P Campbell, Dr H Mackey, T Phelan, Terence Doyle, J J Dolan,

William Toomey, Joseph Bowden, S McGlouglin, C Macauley, T J D Lane, Mrs M Mulvey and C Geoghegan.

The nineteen members of the previous joint committee were: Major T W Kirkwood, W T Cosgrave, A F Buckley, Lt Col J Verschoyle Campbell, H E Guinness, Joseph McGrath, Stephen McGlouglin, Mrs M Mulvey, Patrick J Munden, P J O'Connor, G Brock, Senator S P Campbell, W Boydell, Dermot Findlater, T J D Lane, C J U Murphy, Dr Angela Russell, Henry Stokes and J S Quin. (Public representatives, elected by their own bodies, were Councillors William Rollins, Martin Molloy and Eamon Rooney, and P A Brady, representing Dublin Corporation.) The ten Protestant members of the old committee, seven lay and three medical, all lost their seats.

It was clear there had been a conspiracy to oust the old joint committee. It became known that 33 annual subscriptions were received on 1 April 1949, three days before the triennial meeting, by Mr J I Fitzpatrick, the secretary-manager. These were from people unconnected with the hospital, except for Dr Herbert Mackey. This successful coup was reported widely in the press and seized upon by northern papers, such as the *Londonderry Sentinel* and the *Northern Whig*, with headlines such as 'Eire Board purged of Protestants'.

A special meeting of the new joint committee was held on 9 April 1949 and Dr H O Mackey was elected chairman by nine votes to eight, the former members of the old joint committee and the three county councillors voting against him. Senator Michael Colgan was elected vice-chairman. W T Cosgrave asked the chairman whether there was any complaint against the manner in which the outgoing joint committee had conducted the affairs of the hospital and the reply was in the negative. It was now clear that Dr Mackey was the leader of the group which had succeeded in forming the new joint committee, hence referred to as the 'Mackey joint committee', to avoid confusion.

At a meeting of Dublin County Council on 11 April 1949, the take-over at the Meath Hospital was discussed. Eamon Rooney TD said that most of the new elected members and the persons who voted for them had never attended a general meeting of the Meath Hospital until 4 April, 'therefore it is an organised piece of strategy, and it is not unreasonable to ask why these people became public spirited overnight'. Mrs Mulvey said that she had been a governor of the board of the hospital for fifteen years, but when the chairman and vice-chairman had declined to make a statement on the position which had arisen, she, Mr Cosgrave, Dr Russell and Senator Campbell had resigned from the board.[95] These resignations from the Mackey joint committee were followed by those of Tom Lane, the Revd W Kealy and Mr S McGloughlin. In all, seven Catholics resigned from the new joint committee in protest. Later the entire ladies' committee of the Meath Hospital resigned.

We regret very much having to take this step, but, owing to recent happenings at the annual general meeting when nominees, many of whom had given long honourable service to the hospital, were removed from the board, we have no other alternative.[96]

The Knights of St Columbanus
There is no doubt that the takeover of the Meath in 1949 originated with the Knights of St Columbanus, and that 'action' had been contemplated for some years. Evelyn Bolster (Sister M Angela), in her 1979 history of the Knights of St Columbanus, had access to the minute books and records of the Order. As regards the Meath, she states:

> This hospital was listed as non-denominational but it was Protestant administered and its Appointments Board had the reputation of discriminating against Catholic applicants. Available records indicate that the Supreme Council had the hospital under observation since 1943 when an investigation ordered by Dr Stafford Johnson brought to light the following details supplied on 2 February 1994 by Dr Herbert O Mackey, FRCS, Grand Knight of Dublin Council No 6.

Dr Mackey's report included the following:

> 'In spite of this [system of filling vacancies], some three or four years ago there were three Catholics on the staff: Dr Lane, Dr Oliver Gogarty, and Dr Paul Carton—Dr Gogarty told me a few years ago that he was appointed only after the Dublin Corporation Nominees (on the Board) threatened to recommend the Corporation to withhold the grant if the Hospital proceeded to appoint a young Protestant doctor, just six months qualified: Dr Gogarty himself being a Fellow of the College of Surgeons who had pursued post-graduate courses in nose and throat work in Vienna and had been assistant to Sir Robert Woods at the Richmond for some years.
>
> 'But Dr Gogarty resigned from the Meath Hospital where his position had been rendered untenable by the Board who appointed Dr Furlong from the Adelaide Hospital as his assistant and entrusted to him the teaching of students and postgraduates within the hospital. Dr Paul Carton died. Dr Lane (the one remaining Catholic) is now in a difficult position and considerable pressure is being brought to bear on him. If he resigns the Hospital will be one hundred per cent Protestant . . . I suggest

that the existing system in the Meath Hospital be challenged and fought out, that vacancies be advertised or, at the very least, that appointments should be made openly by the Board of Governors.'

The Supreme Council was non-committal. The hospital was protected by charter and it would be 'very difficult to remedy the present situation' so a policy of non-interference was adopted.

In the aftermath of this enquiry a small unauthorised group of knights under Senator Michael Colgan secured a monopoly on the Joint Appointments Committee of the Meath Hospital.[97]

After the publication of this history of the Order and a radio programme about the book, Professor R E Steen, a former physician to the Meath, wrote a letter to the *Irish Times* on 3 October 1979 in which he stated:

> Though she [Sister Angela] clearly has done a most sincere and conscientious research through the records of the society, I cannot agree that at the time I remember Oliver St John Gogarty on the Meath as ENT surgeon, nor at any time since, has there ever been any sectarian atmosphere whatsoever about appointments to the staff or anything else. Rather than, from one's impression of the broadcast that Gogarty was as it were squeezed out, on the contrary he was a most popular person, and when to the regret of the staff he voluntarily retired gave to the hospital for the boardroom a very lovely antique table round which board meetings are still held, and round which at one time no doubt sat such similar brilliant and colourful personalities as W B Yeats, A E Russell and the like. I was fortunate that Gogarty liked me because he did not spare those he disliked, and his references to the Knights of Columbanus were unprintable.[98]

Certainly the hospital records give no hint of sectarian influences, either at the time of the appointment or at the time of the resignation of Gogarty. At a later meeting of Dublin County Council, Mrs Mulvey further stated:

> I have information that cell six of the Knights of Columbanus was responsible. It is a very poor thing for any organisation, whether Catholic, Freemason or Jewish, or anything else, to set out to try to deprive their fellow men of their livelihood. I know that people were canvassed to pay two guineas to put the 'skids' under those who were on the board. These doctors included Dr Stokes and Mr Lane who have given long, honourable and honorary service to the poor of the city.[99]

Legal Controversy
In May 1949, the members of the old committee initiated legal proceedings against the Mackey joint committee, claiming that the election on 4 April 1949 was irregular and unfair, and should be deemed null and void. There was considerable delay before this came to court, while applications were made to the High Court in November 1949 for the lawyers representing the plaintiffs to have discovery of the documents and records in the hospital, and applications for alternation of statement of claim were made to the High Court in March 1950.

Dr Tom Lane was not optimistic about the legal action and he was advised against it by his friend W T Cosgrave, a former member of the joint committee and ex-Taoiseach and Minister, who said that the best plan was to get a private Bill passed through the Oireachtas reconstituting the governing body of the hospital. He argued that, since the bulk of the finances of the hospital now came from public funds, the governing body should be constituted predominantly from public representatives. A new Act would be needed to bring this about, and he, Mr Cosgrave, would use his influence to initiate this. In March 1950 the text of a private member's Bill about the Meath Hospital, sponsored on an all-party basis by Martin O'Sullivan (Lab), Alderman P S Doyle (FG), M J O'Higgins (FG) and Alderman John McCann (FF), was circulated to members of the Dáil.

Meanwhile, tension developed between the secretary-manager, Mr John Fitzpatrick, and the Mackey joint committee, leading to his dismissal. John Fitzpatrick then made an application to the High Court, in July 1950, for an injunction to restrain the Mackey joint committee from terminating his appointment. The counsel for Mr Fitzpatrick claimed that he had carried out his duties most efficiently and that the present joint committee were 'interlopers and trespassers'. Mr Justice Davitt, before whom the case was presented, rejected the latter reference, as 'that matter was before another court'. He found insufficient grounds to grant the injunction and dismissed the case.[100]

John Fitzpatrick was a founder member of what is now the Irish Health Services Management Institute. A native of Mitchelstown in Co Cork, he has had a distinguished career in accountancy and business and was president of the Federation of Irish Manufacturers from 1957 to 1959. Mr F D Murray was appointed secretary-manager in his place on 17 July 1950. In a recent short biographical account, Fitzpatrick recollected his time at the Meath Hospital:

> Following a rancorous AGM, a totally new governing committee took over. Subsequently, the new committee presented me with a £300 bill for legal advice in relation to an action by their predecessors. They wanted

me to pay it out of hospital funds. When I refused to sign the cheque they told me to leave and accompanied me to the door to see me off the premises . . .

They were interesting times and I look back on them now fondly and with amusement.[101]

The action by the members of the former joint committee finally came before Mr Justice Gavan Duffy on 14 November 1950. The first witness called was the Revd Father William Kealy. He said he had been associated with the hospital for about eight years. At first he had been a visiting priest and later had become a life governor.

On 1 April 1949 Father Kealy had brought thirty-three subscriptions of two guineas to Mr Fitzpatrick, the secretary-manager, from whom he obtained receipts. He knew some of those named on the receipts and probably all of them by repute. He stated that the plan was to elect a new committee and that he had told this to the secretary-manager in confidence and requested him to make no mention of it. The people whose subscriptions he had delivered were also to keep it secret. The reasons he acted thus were because of the reputation of the hospital in the public eye and that it was a place where discrimination was practised against Catholics, though only with regard to medical board appointments. However, he agreed that there was a majority of Catholics on the outgoing board and that, at the general meetings in 1943, 1944 and 1945, he had either proposed or seconded the re-election of the outgoing joint committee.[102] Evidence was also given by Dr Barbara Stokes, Major Kirkwood, Mrs Mary Mulvey, Dr Cecil Robinson and Mr P J Munden, all of whom confirmed the confused state of the election proceedings, as already described.

Having heard the arguments of the counsel for both sides, the president of the High Court dismissed the action of the plaintiffs, the members of the outgoing joint committee. He stated that the law could not interfere with the internal affairs of a company and that the correct procedure in such a dispute was for the plaintiffs to call a general meeting of the governors, in which case there were in the region of up to one hundred members who could exercise their voting rights. He heavily criticised the election of 4 April 1949, describing it as a 'shocking imbroglio'. While expressing sympathy with the plaintiffs, he pointed out that those who voted for the defendants, having paid their subscriptions, were entitled to vote and he noted that some of the plaintiffs who voted had not paid their subscriptions. He further stated:

> The plaintiffs said that the established practice of the corporation for a number of years was to have existing members of the joint committee

reappointed without opposition at the election meetings. They claimed, presumably in point of law, that the practice should not have been broken without due notice to all voters at the April meeting.

Stating that he 'must approach this grievance as a lawyer', the president went on:

> but this ingenious claim is novel to me in law, and I cannot devise a special law of prescription to solace people who have lost an election. As sticklers for propriety, the plaintiffs must remember that former joint committees were hopelessly at sea since 1938 where their legal position was concerned, so that successive joint committees have again and again taken office irregularly since that date.

After the conclusion of the court case, in November 1950, the Mackey joint committee must have felt their position was strengthened and so they proceeded to turn their attention to medical appointments. Professor Jessop received a letter from the Mackey committee asking him whether he wished to continue as a physician or a biochemist to the hospital, as the committee had decided he could not hold both appointments. After failing to reply to this and to two reminders, he was informed that his appointment as biochemist had been terminated. Following this he ceased to receive his salary, although he continued to receive specimens for biochemical estimations in the physiology department of the College of Surgeons, as previously.[103] The Mackey committee also wrote to Dr Deane Oliver, saying that she could not hold two appointments, that of anaesthetist and records supervisor, and that she must resign from one of them. She understandably chose to resign the records post which only carried an honorarium of £50 per annum.

In December 1950 the committee noted that Brandon Stephens, personal assistant to Henry Stokes, and Dr R O'Hanlon, Dr John Cussen and Dr Hickey (assistant to Mr Lane) had never been recommended to the joint committee by the medical board and Mr Stephens was asked what right he had to be carrying out surgical operations in the hospital. The Mackey joint committee then issued advertisements for four posts: two surgeons, one gynaecologist and one pathologist. In February 1951, Thomas O'Neill and Jeremiah Hurley were appointed surgeons and Raymond Cross gynaecologist. Edward Shanahan was appointed assistant surgeon and Dr Dermot O'Kelly pathologist. It was made known to the appointees that, under the new Act, any appointments made after 6 December 1950 by the joint committee could be cancelled. Thomas O'Neill, realising that the appointment was under dispute, did not attend the Meath, but Jeremiah Hurley and Raymond Cross did, which led to unpleasant scenes, as they were appointed to the positions of Montgomery and O'Hanlon who were still working in the hospital.

In preparation for these new appointments the Mackey committee issued new schedules covering bed allocation, operating sessions, out-patient sessions and surgical accident duty. As this was done without the agreement of the existing staff, it was bound to cause confusion and increased tension in the hospital. The medical staff became increasingly apprehensive, realising that the joint committee were determined to change the appointments and the character of the hospital, with consequent unrest and division in the hospital.

Medical board members were advised privately that an approach should be made to the Catholic archbishop of Dublin, John Charles Mc Quaid, and to ask him to intervene. Professor Jessop and Dr Cyril Murphy obtained an appointment with the archbishop, who received them most cordially for tea. Jessop explained the details of the happenings at the hospital and said that he had lost his job and that the Protestants were being 'persecuted'. The archbishop said that he had had nothing to do with the affair and only knew what he had read in the newspapers. However, he said he would do what he could and invited them back two days later. When Jessop and Murphy returned, Archbishop McQuaid said that he had asked Stephen McKenzie, the Head of the Knights of St Columbanus, if the knights had organised the coup, and he was assured that they had not, but that it had been arranged by a small section of members without any authority from the main body. He said that he had then phoned the chairman of the Mackey joint committee but had been unsuccessful in reaching him. He had then sent for the vice-chairman, who had duly arrived, and he felt sure that they would see a result from that interview very soon.[104]

Three members of the Mackey joint committee resigned in February 1950. They were the vice-chairman, Senator Michael Colgan, Joseph Bowden and Charles H Macauley. They wrote a letter to Mr Murray, secretary of the hospital, saying:

> As we pointed out at the meeting, we welcomed the Meath Hospital Bill now before the Oireachtas, as a measure that we believe will achieve, when it becomes law, the objects for which we became members of the Joint Committee in April 1949.
>
> Consequently we felt that it was most injudicious and inequitable that the appointment of four new doctors to the medical staff of the hospital should be proceeded with. We felt that these appointments were, at the moment, superfluous and were bound to intensify the hostility of the

Medical Board to the members of the Joint Committee, without adding in any way to the efficient running of the hospital . . .

It will be agreed that the relations between the Medical Board and the Joint Committee for almost two years now have been deplorable. We are prepared to admit that the Joint Committee were in no way to blame for this. But in our view you cannot run a hospital in the way it should be run without the active co-operation of the medical staff . . . In view of the above and of certain other happenings during the past few months we feel that we cannot identify ourselves with any future decisions of the Joint Committee and that no other course is open to us but to ask you to place our resignations from the Board before the next meeting of the Joint Committee. We are sending a copy of this letter to the press.[105]

Tom Lane became very depressed and for a time was ill in 1949 and 1950, after the take-over of the hospital by the Mackey committee. Following his resignation from that committee, he believed that he had lost any support from its members for his new genito-urinary department, and he even contemplated leaving Dublin. He wrote to Dermot O'Flynn, who was a surgical registrar in urology in Edinburgh, on 25 November 1950:

I know you have been following the Meath Hospital case. I am afraid all chance of a genito-urinary department for me has now gone and with it the hope I had of having you as my assistant.

The news of the major upset in the Meath Hospital and of the possibility that the new GU unit would not be built spread through the urological world internationally. On 8 January 1951, Walter Galbraith, president of the British Association of Urological Surgeons, and David Band, the editor of the British Journal of Urology, wrote a joint letter which was sent to Noel Browne (Minister of Health), Mr Costello (Taoiseach) and Eamonn de Valera, and to the chairman of the Meath Hospital. The following are extracts from that letter:

It has come to our knowledge that the future of the Urological Department of the Meath Hospital, Dublin, under Mr T J D Lane, FRCSI, is threatened. In view of this we thought it might not be regarded as interference if we expressed to you our views on this matter . . .

The department of urology in the Meath Hospital in Dublin under Mr Lane is the only such specialised clinic in Eire; it is generally acknowledged to be one of the most famous in the British Isles and for some years it has been a place of pilgrimage for surgeons coming from abroad. Mr Lane enjoys an unrivalled reputation in the field of surgery in which he has specialised, and the department he has created in the Meath Hospital and the team of assistants he has trained to help him show his genius . . .

We feel that any restriction imposed on him in the development of his urological clinic would be a disaster to the advance of urological specialist practice in the English-speaking world. We trust that some means may be found to ensure the continuance of Mr Lane's urological organisation.

This letter, coming from distinguished medical authorities outside the country, probably had some influence in government circles and assisted the passage of the Meath Hospital Act, which was finally passed through the Dáil in February 1951. It came before the Senate in March and was passed without delay. During the Senate debate, Senator Colgan declared that he:

accepted the bill and wanted to pay tribute to the committee of management elected in 1949. No body could have worked harder in the interests of the hospital, in spite of opposition of the medical staff. No graduates of the National University could get a position there; almost all appointed were from Trinity College, and of eight on the medical staff seven were non-Catholics.

Although the *Irish Times* stated that there was no bar to National University men, the fact remains that they did not get in. There was a little coterie of doctors there who appointed doctors, and actually did appoint their friends and relations.

I had to sit in a board room with a picture of the Famine Queen smiling benignly down on us as we transacted the affairs of the hospital. We have changed that today, there is a crucifix there. That will give you an idea of the change in atmosphere.[106]

Chapter Fourteen

Back to Normal, 1951–2

The Meath Hospital Act of 1951 removed the members of the Mackey joint committee from office and put the Dublin city manager in charge of the hospital until a new committee was elected. The Act confirmed the constitution of the medical board but removed its legal power of appointment of physicians and surgeons which had survived ever since the 1774 Act and the 1815 Act under clause 7. That ancient Act did not exclude graduates of the National University, as is sometimes stated, as this had not yet come into existence, but it did lay down that 'elections shall be from members or licentiates of the King's and Queen's College of Physicians, and Royal College of Surgeons in Ireland'. In practical terms this meant that, as well as their university degree, they had to have membership of the College of Physicians or fellowship of the College of Surgeons, either of which would be essential anyway for a senior hospital appointment.

The new 1951 Act stipulated, under clause 3, that when a vacancy occurs in the medical board this should be advertised, that the medical board should examine the credentials of the applicants and that, following consideration of the report from the medical board, the joint committee should elect the physician or surgeon to fill the vacancy. It would no longer be the case that the medical board elected the candidate and simply informed the committee of governors of their decision.

The composition of the joint committee in future was laid down as follows:

Dublin County Council	5
General Council of County Councils	2
Dublin Corporation	6
Medical Board	4
Hospital Corporation (governors and governesses)	4
Members co-opted by the above elected members	2

Election of the governors would be by proportional representation, by postal ballot. There was also a clause allowing for removal from office of anyone appointed after 6 December 1950. This allowed for dismissal of the controversial appointments made by the Mackey joint committee while the Act was going through the Oireachtas.

The Mackey committee made a long statement protesting against the Act, finding it 'vindictive' and stating that the legislation was carried out with undue haste without a full public enquiry into the hospital's affairs. However, the hospital corporation members were elected in May 1951 to the new joint committee by a strictly controlled postal ballot, set up as directed under the new Act. The four governors elected were previous members of the old joint committee before the coup. They were Gabriel Brock, a principal of the accountants Craig Gardiner, Henry Guinness of the banking firm Guinness and Mahon, Dr Angela Russell and P J O'Connor. The four representatives of the medical board who were elected were Dr Cyril Murphy, Mr T J D Lane, Professor W J Jessop and Dr James Quin. Dublin Corporation members were John McCann TD, Peadar Doyle TD, Walter Breathnach and Dr ffrench O'Carroll TD. Dublin County Council members were Mrs M Mulvey, H P Dockrell TD, W Rollins, J J Clare and Eamon Rooney TD. The General Council of County Councils was represented by Thomas Condon (Meath) and Christopher Byrne (Wicklow). Alderman John McCann was elected chairman, and councillor Mrs M Mulvey was vice-chairman.

In May 1951, during the interregnum while the city manager was in control of the hospital, a very unfortunate incident occurred related to the hostile atmosphere between the medical board and the extra surgical appointments made by the outgoing joint committee. The medical board members, the house staff and students were all opposed to the four surgical appointments which had been made by the now defunct joint committee while the Meath Hospital Act was going through the Dáil. The nursing staff were also sympathetic to the existing staff, but the matron, Miss Magee, had warned the nurses 'don't get involved—look after the patients'.

At this time, Mr Jeremiah Hurley made a determined effort to get a bed in the hospital for a patient referred to him whom he had diagnosed as having intestinal obstruction and requiring urgent surgery. He was told by the house staff that there were no beds, and so he spoke to Mr Lane and Mr Stokes, and also the city manager at City Hall, who were all unable to get him a bed. It was well known that the medical staff were not prepared to co-operate with the new appointees and tried to prevent them having access to beds. The patient was eventually admitted, after considerable delay, to the private hospital, the Bons Secours in Glasnevin, where the patient died following operation. An inquest was held, lasting for twelve hours, with Mr Lane and Mr Stokes called to give evidence, as well as P J Hernon, the city manager. The jury issued a rider 'condemning the action of the Meath Hospital Board in denying Dr Hurley, who was a member of the staff, facilities to operate on the patient'.[107] This serious incident received wide publicity and there was adverse comment in the press about the behaviour of the doctors at the Meath Hospital. It was a very unfortunate result of a highly abnormal situation of conflict in the hospital.

Return to Normality
Under the 1951 Act, the hospital finally settled down to a peaceful course. There was something for all sides in the Act and it seemed a just solution to the serious dispute in the hospital. It gave substantial representation to public representatives, the governors had four representatives with well-defined arrangements for election procedures, the medical board had four representatives, as they had had before, and the Catholic activists saw that the medical staff could no longer be elected and appointed solely by the medical board but that this was now in the hands of the joint committee, on the advice of the medical board.

The new joint committee reorganised some of the medical appointments. First of all, the appointments made by the committee under Dr Mackey were all rescinded, as provided for in the 1951 Act. On the advice of the medical board, R F J Henry ceased to be reserve surgeon and was made consulting thoracic surgeon. In November 1951 the joint committee appointed Brandon Stephens assistant surgeon and Roderick O'Hanlon assistant gynaecologist, both of whom had previously been personal assistants. In October 1952 the medical board recommended to the joint committee that Victor Lane and Dermot O'Flynn be appointed assistant surgeons to the urological department.

In November 1952, according to the new procedure, the post of senior surgeon was advertised. Four applications were received. These were assessed and graded and the results sent to the joint committee. Mr Douglas Montgomery, who had been reserve surgeon for some years, was appointed by the joint committee as full surgeon. Montgomery was born in the USA and had two passports. He graduated in medicine at Trinity College in 1940 and served in the RAMC during the war. He was at the Normandy landing and was said to have been the first surgeon to operate on a battle casualty on the beachhead in France on D-day. In jumping ashore from the boat he injured his back, which caused him to be invalided back to Ireland, but when he returned to Dublin he found that he had lost his opportunity for a senior post at Dun's Hospital, where he had trained. From the time he first became associated with the Meath Hospital in 1945, he developed a reputation for medico-legal work. He was a meticulous man, even using shorthand when interviewing patients. He was also ambidextrous, which was quite impressive when performing operations. He was enthusiastic in the use of the peritoneoscope, which was frowned on by his contemporaries but which heralded the endoscopy of today. He had a rather introverted military manner, and was well suited in his role as commissioner of the St John's Ambulance Brigade, to which he gave excellent service. He was also active in recommending preventive measures against traffic accidents. Montgomery was to die in office in 1974.

New Nurses Home

While the Mackey joint committee had been in office, the building of the new nurses' home had proceeded and this was expected to be opened in April 1952. The original plan had been for the old nurses' home to be converted into the urological department (GU unit), but it was subsequently agreed that the west wing should be razed and that this should be replaced by the GU unit, with a stipulation that the front stonework of the west wing be preserved. On 5 November 1951 the works committee reported to the joint committee that there were three major projects planned:

- A. Conversion of the old nurses' home for general accommodation, £20,000.
- B. Extension of the kitchen, dining room and ancillary accommodation, £35,000.
- C. Erection of GU unit on site of west wing, £95,000.

However, in April 1952 a revised estimate for the demolition of the west wing and construction of the GU unit was calculated to be far in excess of the Department of Health grant of £125,000, so the committee chose instead to convert the old nurses' home for the GU department. This would require the addition of two storeys and a lateral extension towards the drive, costing £90,000, which would leave £30,000 for ancillary work.

Finally, on 19 May 1952, at a special meeting of the joint committee, it was decided to demolish the old nurses' home and build a 'completely new structure faced in redbrick work'. It was rumoured later that this radical decision was taken because of dry rot being discovered in the old nurses' home. Though this was not mentioned in the minutes of this meeting, dry rot had been reported by Robinson, the architect, at a meeting of the works committee in December 1951. At that meeting it was agreed 'to postpone following the dry rot to its source until the nurses' home has been vacated'.

The decision to demolish the nurses' home came as a surprise to some of the members of the medical board, who considered that if any building was to be removed it should be the west wing. A special meeting of the medical board was held in June 1952 to consider the situation. There was a full attendance, including Mr Lane. Following a protracted discussion, during which Mr Stokes walked out in protest, it was decided to send a letter to the joint committee:

> The Medical Board has learned with satisfaction that the Minister of Health has agreed to allow the scheme for the establishment of the urological unit to go forward. The Board however is sorry that it has not been found possible to include the reconstruction of the West Wing in the present plans as that building has many structural disadvantages as a ward block. The Board would ask the Joint Committee to bear this point in mind for future development.

The opposition of the majority of the medical board to the demolition of the old nurses' home building must have irritated Lane, who felt that they were being obstructive towards the establishment of a urological unit. From that time there was some coolness between Lane and his colleagues, and he only attended one medical board meeting again, in October 1952, though he remained a representative of the medical board on the joint committee.

Meanwhile, the new nurses' home was opened in March 1952 by Dr James Ryan, the Minister for Health. It had cost £130,000. It provided modern accommodation for 120 sisters and nurses, with a lecture room and a recreation area. It made some space available in the main hospital and the west wing as bedrooms occupied by the sisters became available.

T J D Lane's reputation was now such that patients were coming to him from all over the country. There was an overflow of his patients into all the surgical beds which would not be relieved until the new GU unit was constructed in 1955. There were 825 admissions to the urological department (which included beds at St Kevin's Hospital) for the year 1952–3. Of these, 217 were prostatectomies, with only 4 deaths, although twenty years previously the mortality for prostatectomy at the Meath Hospital had been 20 per cent.[108]

When no beds were available for ambulant patients from outside of Dublin who needed investigation, they were accommodated in 'digs' in Heytesbury Street and even, when there was an overflow, in the Iveagh Hostel. This caused anxiety to the joint committee, from the legal point of view, in having hospital patients boarded out and not directly under the care of the hospital staff.

In those days there were many cases of tuberculosis of the urogenital tract, which required prolonged treatment and observation. In November 1952 a building which became known as St Catherine's Annexe in Pigeon House Road was rented from Dublin Corporation for patients with urological tuberculosis. It had eighteen beds and it was financed by the efforts of the ladies' committee.

Sadly, Dr Harold Pringle died in 1952 at a relatively young age. As the first specialist radiologist appointed to the hospital he had been struggling to expand and consolidate the X-ray service. He came from a well-known medical family and was very popular with his colleagues. Miss Lucy Dimond, who had been sister on the 'medical landing' for thirty years, also died in 1952. Her bedroom on the medical landing became the 'Dimond ward' in her memory and was given over to Dr Cyril Murphy for his allergy work. He used it to isolate patients from potential allergens and was able to identify sensitising agents which would precipitate allergic attacks, such as asthma.

Chapter Fifteen

The New Genito-Urinary Unit, 1953–60

A significant year for medicine in Ireland was the year of 1953. The appointed day for the introduction of the compulsory intern year was 1 January. Before this, each Dublin Hospital had had a few qualified doctors as house physicians and surgeons, and the rest of the resident staff were students. Now all newly qualified doctors had to do six months of medicine and six months of surgery in residence before they could obtain their final qualification to practise. The Meath was able to accommodate eleven interns, but this left little room for resident students, which had been a traditional form of training in Irish medicine. The other major impact was the 1953 Health Act, which the medical profession and the Catholic hierarchy had opposed, on the grounds that the state was taking too much control over medical practice.[109]

One aspect of this legislation was that the local authority was to provide payment to the consulting staff of voluntary hospitals, who, up until then, had received no payment for treating public patients in the wards. A system was devised in which the local authority paid a sum of money into a 'doctors' pool' in the hospital, the amount dependent on the number of occupied beds. Initially, two shillings and sixpence per day per occupied bed was paid. The consultant staff divided the accrued sum (known as the ISA fund) amongst themselves on a unit system: a surgeon received twenty units, a physician eighteen and a gynaecologist five units. In 1957 the total pool amounted to £4,500 and this was divided into 137 units. This extraordinary system worked

contd on page 153

Staff assembled for Queen Victoria's visit in April 1900.
Front row, sixth from left and onwards: William Taylor, surgeon; Dr Edward Lennon; Dr John W Moore; Dr James Craig; Francis Penrose, hospital registrar; Richard Lane Joynt, surgeon.

The Red Cross Nursing School and medical staff in 1902. Some notable figures in this picture include: Standing on left, third row: Surgeon Richard Lane Joynt. Seated on left, second row: Dr Edward Lennon. Centre, with dog: Ellinor Lyons, matron. Seated second from right, second row: Dr James Craig. Standing on right, third row: Tenison Lyons, apothecary. Standing on right, back row: Surgeon William Taylor.

Nursing staff in 1924.
Seated (second from left and onwards) H Stokes, W Boxwell, Sir J Moore, Matron Wall, Secretary Dow, T J D Lane, C W Murphy, unknown.

Medical staff in 1927.

(Left to right) Front row: P Aird, Jim Ryan, M F Dunphy, I LeRoux.
Middle row: W Boxwell, E Lennon, Sir John Moore, R Lane Joynt, O St John Gogarty, P Carton, H Stokes.
Back row: C J Murphy, D S Austin, J A K Douglas, D J Keane, J Molony, W Alexander Roberts.

A tinfoil collection in aid of the hospital during the 1920s.

Workers removing a car which had been placed in the entrance hall by students in 1958.

Mock-up party using the busts of 'the greats', arranged by the students in 1937.

Oliver St John Gogarty presented this sculpture of himself to the hospital in 1938.

Henry Stokes, surgeon 1912-1955.

Sir John W Moore, arriving at the hospital in 1936, aged 91.

Professor W J E Jessop, biochemist and physician 1930-1980.

Nurses' Conference, c.1938.
(*Left to right*) *Front row*: Cecil Robinson, unknown, Henry Stokes, Matron W Gage, unknown, Cyril Murphy, Margaret Lytle, James Quin.

Maureen Murphy, anaesthetist 1940-1962, and medical missionary 1962-1991.

Brendan O'Brien, physician 1943-1974 (from a portrait by J G le Jeune).

Dermot O'Flynn, GU surgeon 1952-1987.

Victor Lane, GU surgeon 1952-1990.

The official opening of the Genito-urinary Unit on 17 November 1955. (*Left to right*) Minister of Health, T F O'Higgins, T J D Lane, Councillor Mrs Mulvey, Sister Betty Cunningham.

Bert Keating, porter 1913-1963.

Dr Cyril Murphy with the Stokes loving cup at the annual dinner of 1955.

David Lane, surgeon (and musician) 1957-1971.

Staff and students in 1960.

(Left to right) *Front row:* W J Jessop, B Stephens, C Robinson, D Montgomery, B Mayne, P Gatenby, J Quin.
Second row: S Douglas, D Robinson, V Keeley, M McKay, J Jessop, D Weir, G Edwards.
Third row: F Penco, R Sinniah, I J Temperley.
Back row: R O'Hanlon, B Keating (porter), J Gummer, D Waldron-Lynch, C Comer, H Khonsari, unknown, Sinniah.

T J D Lane sharing a joke with Matron Ann Magee in 1965.

Elizabeth O'Dwyer, matron 1968-1996.

Sisters of the GU department in 1980.
(Left to right) Maura Dunne, Eileen Sheridan, Marie Cooney, Maureen Fallon.

Consultant medical staff, 1983.
(Left to right) Front row: C O'Morain, A Tanner, S Hamilton, F Keane, D McInerney.
Middle row: V Lane, S Douglas, M O'Callaghan, B Mayne, D O'Flynn, B Stephens, B Brennan, M Butler, G Hurley.
Back row: J Barragry, M Pegum, M Cullen, K O'Sullivan, D Robinson, J Fitzpatrick, G Mullett, W Beesley, F O'Loughran, B Keogh, J Galvin, G Owens.

DUBLIN'S MEATH HOSPITAL 151

Nursing School in 1994.
Front row: Mary Briscoe, Angela O'Donoghue, Margaret McCarthy, Michael Colgan, Elizabeth O'Dwyer (matron), Mary Cotter, Liam Plunkett, Mary Lennon, Angela Hoey.

Aerial photograph taken in May 1996 showing the construction of the new hospital at Tallaght in south-west Dublin. The Meath Hospital is due to relocate here in 1997. *(courtesy of Tallaght Regional Hospital Board)*

contd from page 136

reasonably well for many years and lasted until the 1980s. Payment was also made by the local authority for out-patient sessions, which was related to the number of patients attending.

The cost of running the hospital was steadily mounting and the expenses for 1953 were £110,844 and the income £52,732, leaving a deficit of £58,112. Voluntary subscriptions amounted only to £1,176.

Opening of the Genito-Urinary Unit, 1955

A major event in the history of the Meath Hospital was the opening of the genito-urinary unit, on 17 November 1955, by the Minister for Health Mr T F O'Higgins. Built adjoining the main building, it was a self-contained four-storey block. As a reflection perhaps of Tom Lane's defensive psychology, it was planned to be completely walled off from the main building. Initially, the only corridor connection with the main building was at ground level. This was essential to allow direct access to the X-ray department, and also to permit food to be supplied from the main kitchen. The department contained eighty beds, the wards arranged in units of six beds, four beds, two beds and single rooms. There was a treatment room, a day room and service kitchen on each floor and the unit had its own operating suite and out-patient clinic. The total cost of the new building was £181,000, which included some improvements to the kitchen.

The urological work was now conducted under dramatically improved conditions. The surgical team was led by Tom Lane, assisted by Dermot O'Flynn and Victor Lane (son of Tom Lane), and a separate panel of anaesthetists had been appointed—C H (Bertie) Wilson, John Cussen and Michael McGrath. The GU unit was greatly admired by visitors from near and far, and was considered to be one of the best specialist surgical units in Britain or Ireland. As an eighty-bed urological unit, it was the biggest in these islands and the second largest in Europe. A distinguished American visitor, Nobel prizewinner Charles Huggins of Chicago, wrote to the joint committee suggesting that the unit be named the 'Lane Unit', but this idea was rejected by Tom Lane himself.

The expansion of urological work increased the load on the X-ray department and pathology services. Dr Sholto Douglas, who had been appointed radiologist in succession to Dr Pringle in 1953, fought hard to get better X-ray equipment and an increase of staff to meet the demand. Laboratory services were provided by departments in Trinity College, biochemistry by Professor Jessop, pathology by Professor O'Meara and his assistants, and bacteriology by Professor F S Stewart. There were also special demands for expert histological opinion in urology and for parathyroid cases, and Dr Joan Mullaney from Trinity College was appointed pathologist in 1956. It was in this year also that the posts of medical and surgical registrars were created for the first time at the Meath.

The transfer of urological beds from the main building freed surgical beds there and so it became necessary to appoint another general surgeon, in addition to Douglas Montgomery, as Henry Stokes was due to retire. The post was advertised and there were three applicants and, in April 1955, the joint committee appointed E Brandon Stephens, on the advice of the medical board, as he had been assistant surgeon for five years and was very familiar with the hospital. The general surgical side was further strengthened by the appointment of Derek Robinson as clinical assistant in surgery. Robinson had been personal assistant to Montgomery since 1953. Later he was appointed assistant surgeon and, in 1972, he became consultant accident surgeon until his retirement in 1989. He succeeded Douglas Montgomery as commissioner of the St John's Ambulance Brigade of Ireland in 1974.

Hospital Staff from 1956
Henry Stokes resigned in June 1956, having served the hospital for 52 years. He had also been surgeon to Cork Street Fever Hospital and the Children's Hospital. He was the last of the famous Stokes family to be on the staff of the Meath Hospital. The medical board and the joint committee agreed that, following his resignation, he be made honorary consulting surgeon to the hospital. His colleagues made him a presentation of an illuminated address, which included the following:

> For almost one hundred and fifty years there has been a close connection between the members of the Stokes family, but of all the distinguished doctors who have helped in the making of that tradition, no one has been more loyal and devoted to the interests of the hospital than the man whom we honour today.
> ... Henry Stokes has continued, extended and added fresh lustre to an already great tradition of service in the field of medicine to Ireland and to humanity. His patients have found in him not only a surgeon whose touch brings healing but also a kindly and beloved physician. To his pupils he has shown himself as a revered and inspiring teacher and a wise and sympathetic counsellor.

Henry Stokes, who was extremely kind to his patients, is also remembered as something of a 'character'. Brandon Stephens, who had been his personal assistant, related the following memories of him.

While he was consulting surgeon to Cork Street Fever Hospital, operations had to be performed there in the bed, as there was no operating theatre. The patients could not be transferred to the Meath as they were infectious fever cases, so the equipment and instruments were brought over from the Meath in sterile towels. Dr Deane Oliver gave the anaesthetic and the operation was done with the aid of an anglepoise lamp.

For the treatment of haemorrhoids, Henry Stokes would use cautery with a red-hot poker heated on a gas ring, while attired in his shirtsleeves and waistcoat. The haemorrhoids were clamped and then cautery applied. It is said that the patients had no pain afterwards. He was also one of the first to do parathyroid surgery in Ireland. The patients were referred from Lane's unit, where these cases presented with renal stones. Accurate blood calcium estimations by Jessop aided recognition and treatment.

In his private life, Stokes was an expert at finding skeletons of the extinct Irish elk. He was able to recognise promising places to excavate, near water where they would have gathered to drink. For a while he had the antlers of an elk exhibited in the front hall of the hospital. He was clearly not impressed by modern communication, using the stump of a pencil to make notes. An irreverent remark of his made about the new GU unit was 'one hormone could close that place'!

After inspections by American authorities and the General Medical Council, the medical schools had come under criticism for their lack of organisation in clinical teaching and pathology. One consequence of this was the decision to appoint medical and surgical tutors in the hospitals jointly between the medical school and the hospital. So, in 1956, when David Lane was appointed assistant surgeon to the Meath to fill the gap left by Mr Stephens' promotion, he was also made surgical tutor for Trinity College.

Cyril Murphy, the senior physician, died on 6 January 1957, at the age of 57. Though he had been suffering for some time with a dreadful carcinoma of the thyroid, he went on with his work, without complaint, until shortly before his death. He had been associated with the Meath all his professional life. After qualification at Trinity, he had secured the medical travelling prize, which enabled him to work at the Mayo Clinic. He was the youngest son of the Revd J E H Murphy, who was rector in Enfield, Co Meath, and professor of Irish at Trinity College. His mother was Emmeline Lennon, a sister of Edward Lennon, former physician to the Meath. Cyril was an upright, conscientious person of deep convictions and a prominent member of his church. He was a loyal servant of the hospital, and honorary secretary of the medical board for ten years. He has been described as possessing 'irrepressible cheerfulness' and his extrovert personality was an important ingredient in his clinical teaching, where he was apt to give a 'picturesque exposition of clinical signs'. One remembers his oral reproduction of the auscultatory signs of mitral stenosis uttered in full voice!

He was married to Maureen Dorman, who, together with Dr Deane Oliver, had been one of the original specialist anaesthetists to the hospital. In 1959 she gave £1,000 to the medical board to set up a lending library for the interns and students, in memory of Cyril Murphy. This was the foundation of the present combined medical and nursing library. In 1961 Maureen resigned from the hospital and went to India as a medical missionary. A quiet, dedicated person, she gave great service firstly at the Christian Medical College in Ludhiana and then at St Columba's Hospital, Hazaribagh. She retired to Ireland in 1982 and died in 1991.

In April 1957, the vacancy created by the death of Cyril Murphy was advertised. A number of applications were received, and I was appointed by the joint committee, on the advice of the medical board.

Towards the end of the 1950s, the medical staff were as follows:

Physicians	W J Jessop, Brendan O'Brien, Brian Mayne, Peter Gatenby
Honorary consulting surgeon	Henry Stokes
General surgeons	Douglas Montgomery, Brandon Stephens, and assistant David Lane
Urological surgeons	Tom Lane, and assistants Dermot O'Flynn and Victor Lane
Clinical assistant in surgery	Derek Robinson
Gynaecologists	James Quin, with assistant Roderick O'Hanlon
Laryngologist	Sydney Furlong
Anaesthetists	
	General—S Deane Oliver, Maureen Murphy, Patricia Delaney
	Urological—C H Wilson, John Cussen, Michael McGrath, Fergus Quilty, Fiona Acheson
Radiologists	Sholto Douglas, with assistant Joan MacCarthy
Alienist	J N P Moore

Ophthalmic surgeon	T J Macdougald
Dermatologist	Augusta Young
Pathologist	Joan Mullaney

The hospital secretary, F D Murray, resigned in 1954 to study for the priesthood. John Griffin was appointed in his place but did not stay long. Mr John M Colfer was appointed in 1957 and was to stay for 25 years, giving excellent service. About this time the GU unit was shocked to lose two young members of its staff. Anaesthetist John Cussen, with a popular extrovert personality, died suddenly of coronary disease at the age of 42 in May 1959. In March 1960, clinical assistant Dr John Phelan was killed in a traffic accident. Dr Kevin P O'Sullivan was appointed anaesthetist to the GU department in 1960.

Federation and Amalgamation Legislation

During 1958 and 1959 the small Dublin hospitals had informal discussions about amalgamation, actively supported by the Department of Health. An Act passed through the Dáil and the Senate in 1960 which established federation, enabling eventual amalgamation. This became law in 1961.

There were seven hospitals involved and, apart from the Meath, these were the Adelaide, Harcourt Street Children's Hospital, Dr Steevens', Dun's, Mercer's and Baggot Street. A central council was formed, with five representatives—three lay and two medical—from each of the hospitals. This gave a rather unwieldy council of thirty-five people and there were also additional representatives from the medical schools. The council had important basic powers of appointments and approving capital expenditure, and agreeing on the location of specialist units. In 1961, under the Hospitals Federation and Amalgamation Act, the governing body of the hospital, known up to then as the 'joint committee', became the 'hospital board'. Also, under the Health Authority Act of 1960, the authority of Dublin Corporation and County Council to elect members to the joint committee was transferred to the Dublin Health Authority and later to the Eastern Health Board.

Chapter Sixteen

Medical Specialisation, 1961–72

In 1960, as a result of adverse reports on Irish medical schools, Trinity College decided to set up full-time professorial units for the clinical subjects. Up until then, the professors of medicine, surgery and obstetrics in the medical schools of the Republic were all part-time, and dependent on private practice. I was appointed professor of clinical medicine, and Robert Steen, formerly at the Meath, was appointed professor of paediatrics at Harcourt Street Children's Hospital.

My task was to develop a full-time professorial unit at the Meath Hospital. The initial problem was space. For the first year, Sydney Furlong kindly allowed me to use the out-patient room which I used for my ear, nose and throat clinic two days a week. This was the very room that Gogarty had used for his dispensary, still with its separate escape door. Donald Weir, who was medical registrar, became the first assistant in the unit. He had trained in gastroenterology in Edinburgh and returned to engage in clinical research in association with Ian Temperley, haematologist and lecturer in pathology in Trinity.

With the increased activity in teaching and research resulting from the new professorial unit at the Meath, there was an urgent need for a lecture/conference room. It was not feasible to build the offices, research laboratory and conference room essential for a professorial unit, because capital expenditure was controlled by the central council of the Federated Hospitals, which would not have approved permanent construction at the hospital site as it had no planned future. Neither was Trinity College prepared to invest a significant amount of money in a hospital with an uncertain future, apart from having financial constraints of its own.

So the only solution was to construct 'temporary buildings', and permission was granted by the joint committee for Trinity to construct a 'Heron' wooden hut behind the west wing to accommodate the professorial unit. This cost £4,000, initially providing two offices and a laboratory, and in 1968 it was doubled in size. It was also agreed that hospital funds would be used to erect a prefabricated 'Bantile' building between the west wing and the main building, on the last remaining piece of grass, to accommodate a lecture room and the physiotherapy department at the north end, next to the west wing. The lecture room, which occupied three-quarters of the building, was formally opened in November 1962. It cost £3,500 and was subscribed to by the medical board, Trinity College, the ladies' committee and the Department of Health. The construction of these buildings enabled the Trinity professorial unit to function reasonably well at the Meath for about fifteen years.

In 1964 Donald Weir was appointed lecturer in medicine by Trinity College and assistant physician at the Meath. There was close association between him and Ian Temperley, bringing together their respective interests in gastroenterology and haematology, which provided a good background for research. Supported by grants from the Medical Research Council, a succession of medical graduates interested in academic medicine worked there in clinical research and obtained postgraduate degrees. They included Connolly Norman, Marcus Webb, Mervyn Taylor, George McDonald, Adrian Fine, Deen Sharma, Owen Morgan and Desmond Sheridan. Owen Morgan is now professor and chairman of the department of medicine at the University of the West Indies in Jamaica, and Desmond Sheridan is professor of cardiology at St Mary's Hospital, London. Donald Weir, Ian Temperley and Marcus Webb all subsequently achieved full professorial status in the medical school of Trinity College.

The deficit of the hospital finances for the year 1956 had now reached £70,000. This was the highest ever and radical measures for economising were discussed by the joint committee, such as closing the convalescent home in Bray and terminating the almoners' appointments. However, these ideas were opposed by the medical board. Maintaining and renovating the old part of the hospital as economically as possible became a constant problem. This was carried out by the maintenance engineer, Tommy Barnett, who was appointed in 1958 and who was a genius at improvisation.

New Faces and Old Friends

Since federation had become law in 1961, appointments were no longer to individual hospitals but to the Federated Hospitals group, thus allowing for re-assignments in case of future amalgamation. Peter Morck was appointed part-time anaesthetist in 1961, after the departure of Maureen Murphy, and Hugh J (Joe) Galvin became full-time senior anaesthetist in 1962. Peter showed his considerable expertise in converting the Smyly children's ward into an excellent intensive care ward. Joe, being full-time, was able to give all his energy to his hospital work, both technically and in committee work. He was honorary secretary of the medical board, from April 1966, for many years.

In 1963 Dermot O'Flynn and Victor Lane were appointed senior surgeons to the Federated Dublin Voluntary Hospitals and assigned to the Meath Hospital, where they became members of the medical board. They made a unique partnership and worked well together for many years, their personalities complementing one another. Dermot was a steady, calm person and Victor, son of Tom Lane, was a warm, friendly personality, who avoided controversy and had a limited tolerance for tense committee meetings! Dr Joan Mullaney, who was reader in pathology at Trinity College, resigned from Trinity but retained her appointment at the Meath Hospital as pathologist. This gave rise to administrative problems between the Meath Hospital and Trinity College regarding pathology services. This was further complicated by the Federated Hospitals system, which was endeavouring to unite the specialist pathology services for the seven hospitals. Discussions on pathology services went on for some years between representatives of the Meath, the Federated Hospitals Council and Trinity College. After ten years serving the Meath, Dr Mullaney resigned in 1966 to give all her time to pathology at the Royal Victoria Eye and Ear Hospital, where she had also been serving for some years and had become very distinguished in ophthalmic pathology.

In 1967 David Lane was promoted from assistant surgeon to surgeon to the Federated Hospitals, with assignment to the Meath. He had proved himself to be an excellent surgeon and a very popular teacher, apart from being a distinguished classical musician. Ian Temperley, who was now haematologist for the Federated Hospitals, was also appointed in charge of the laboratory at the Meath, where the haematology services for the federation were to be centralised. Dermot Hourihane, later to become professor of pathology at Trinity, was appointed to carry out the histological work formerly done by Dr Mullaney.

Also in 1967, death removed several prominent figures from the Meath Hospital scene. T J D Lane died in the February. He had done more than anyone to promote the hospital and bring it fame. Though medicine at the

Meath was at its most famous in the nineteenth century, specialised surgery, developed by Lane against considerable odds, was at its height at the Meath in the mid-twentieth century. Contrary to the orthodox route of a surgical career, he began as radiologist. His only postgraduate degree, the DTMH, was in tropical medicine, but he eventually received an honorary MD from Trinity College and honorary fellowship of the College of Surgeons.

It was an uphill struggle to develop specialised surgical urology in the general wards of the old Meath Hospital, and Lane achieved this by the force of his personality and the high standards of his work. Dermot O'Flynn, who assisted him for years, gave this intimate description of him:

> He was an intense and complex personality, hard working and tireless and totally devoted to what he was doing for his patients, and they recognised this and admired him for it. He was a perfectionist with a short temper which erupted whenever standards dropped below his high level. He had a great ability to inspire people to do better and he could extract extraordinary standards of work from all who associated with him. Possibly his greatest asset was his persistence and perseverance and no problem and especially no patient's problem, was allowed to rest until every possible solution had been explored. Failure was accepted with bad grace. He was probably one of the last of the great individualistic surgeons of this country, but he had the vision to see that for the future, specialisation and teamwork were inseparable and he designed his department on these lines . . .
>
> He knew the joys and terrors of cruising in the Irish Sea. He was fond of children and dogs and he had a succession of Yorkshire Terriers all called 'Sam' who all did the same tricks. He was a great gardener and as one would expect, a very scientific one. His garden was furnished with a delightful selection of exotic and unusual shrubs, all bedded and sited in optimum conditions.[110]

In 1968, a bronze plaque in memory of him was formally presented on behalf of the memorial committee, and a lectureship in urology, founded in his memory at Trinity College, was announced at the annual meeting in May 1970 by the Tánaiste and Minister for Health, Erskine Childers.

Councillor Mrs Mary Mulvey died in May 1967, a few days after attending a finance committee meeting. She had given loyal service as a governess of the hospital for twenty-six years. She was co-opted to the joint committee in December 1941 and had been chairman since 1954. She had taken a leading part in the ladies' committee in raising money for many projects for the hospital, including St Mary's Annexe and St Catherine's Annexe, and had negotiated the transfer of the TB clinic from the Dublin Health Authority to the hospital for use as an out-patient building, at no cost to the hospital. Another longstanding member of the joint committee who died in 1967 was Gabriel Brock, who had served on the joint committee from 1944 to 1965. A senior accountant of the firm Craig Gardiner, he had given expert and valuable advice on the financial affairs of the hospital as chairman of the finance committee.

There have been many other lay members of the joint committee who have given selfless voluntary service to the Meath over the years. One such was Patrick J O'Connor, who had been on the joint committee since 1943 and chairman of the works committee for many years when he resigned in 1968. His place was filled by P V Doyle, the well-known hotelier. Henry E Guinness, a director of the bankers Guinness and Mahon, retired in January 1970 after an astonishing 38 years on the joint committee. Henry Stokes and Cecil Robinson both died in 1967. Both were in retirement after long careers on the hospital staff which have already been described.

Gallagher Administrative Building
In 1969 it was decided to build an administrative building between the gate and the out-patient building. This was at the instigation of Mr P V Doyle, who had the area surveyed and undertook to have a two-storey building erected at the very low cost of £20,000. This project was made possible by a generous donation of £10,000 by Mr Matt Gallagher, who represented the Dublin Health Authority on the joint committee. Mr Doyle contributed £2,000 to the building fund and supervised the rapid construction of the building, which provided a boardroom, offices for the secretary-manager and his staff and for medical and administrative records, and an admissions office. This released space in the old building for extra wards and an extended pharmacy. The children's ward (Smyly ward) was moved down to the first floor, thus releasing space for an intensive care unit on the surgical landing.

The intensive care unit was constructed under the direction of Dr Peter Morck with £10,000 received from the executors of the estate of the late Sir Alfred Chester Beatty. This unit was formally opened by the chairman, Robert Graham, in October 1972. Robert Graham was a very effective chairman of the hospital board because of his special knowledge of the medical scene. He held a senior position with the Voluntary Insurance Board, but in 1982 his services to the Meath were lost, as he left Ireland to become deputy chief executive of the British United Provident Association.

CHAPTER SEVENTEEN

Further Specialisation, 1972–92

The Meath Hospital Act of 1951 defined the medical board as consisting of 'the physicians and surgeons of the hospital', but it did not give adequate representation to the rapidly expanding medical staff, the result of advancing specialisation. In 1971 a medical advisory committee was formed, consisting of all the consultant members of staff, and representatives of the registrars and junior medical staff. By 1975 the functions of the original medical board had been reduced to the election of four representatives to the hospital board and management of several financial accounts—funds for student prizes and distribution of the 'pool money' to pay the physicians and surgeons and occasional consultants.

In September 1975, Victor Lane proposed that the medical board and the medical advisory committee should be replaced by a single medical committee which would elect an executive committee every year. The medical board replied that they could see no reason why the management of the financial funds should be handed over to any other body. They pointed out that election of medical consultants to the hospital board was not confined to members of the medical board, as the hospital board had the power to elect additional medical members. They believed it was their duty to preserve for future generations the mementoes which have been presented to the medical board throughout the 200-year history of the hospital which are on display in the old boardroom. They also pointed out that the medical board had a valuable function in acting in a confidential advisory capacity to its junior colleagues and others when matters of a controversial nature arose, thus avoiding publicity that could be detrimental to individuals or the hospital.

The traditional medical board, with its legal status, survived and in 1978 it was agreed that those members of the consultant staff who were employed for five or more sessions per week would be eligible for membership of the medical board. Accordingly, the medical board expanded to include radiologists Sholto Douglas and Gerry Hurley, and anaesthetist Joe Galvin. Despite applications in subsequent years for membership to the medical board from aggrieved medical consultants who were employed for less than five sessions per week, the rule was not changed. During the 1980s the medical administrative scene was further complicated by the formation of joint medical committees and specialised medical committees for the staff of the three hospitals of the 'MANCH' group (Meath, Adelaide and Children's Hospital, Harcourt Street).

Specialised Medicine
The Trinity department of clinical medicine at the Meath was responsible for fostering specialised branches of medicine. In association with this unit, Donald Weir had been carrying out gastroenterological research and Ian Temperley had been appointed haematologist and had created a haematology department for the Federated Hospitals. In 1967 Donald Weir was appointed physician at Sir Patrick Dun's Hospital by the Federated Hospitals, where he transferred and developed a gastroenterology unit. He subsequently became associate professor of medicine for Trinity College and finally professor and head of the department of medicine centred at St James's Hospital. David Lane, who specialised in gastrointestinal surgery, transferred to Sir Patrick Dun's Hospital in 1971, where he worked in association with the department of gastroenterology which had been set up by Donald Weir. John Michael Pegum, who specialised in orthopaedics, was appointed in his place. In addition to his appointment as surgeon to the Meath, he was lecturer in surgery at Trinity College.

The coronary care ward was started in the 1960s in very unsuitable accommodation in the main building, but this was transferred to the ground floor of the west wing and funds were eventually expended on providing modern facilities, from 1977 onwards.

The National Haemophilia Centre was established at the Meath, in conjunction with Harcourt Street Children's Hospital, and was formally opened by the Minister of Health, Erskine Childers, in 1971. The centre was under the direction of Ian Temperley and was the first such unit in the

Republic. An essential component was the co-operation of the National Blood Transfusion Service, under the helpful direction of Dr Jack O'Riordan. Clinical facilities were provided by the physicians and specialist nurses were trained in the administration of coagulation factors VIII and IX to arrest bleeding episodes. This service meant that haemophiliacs all over the country could, for the first time ever, undergo major surgical operations. Dental extractions and dental care were provided by Gerald Owens and, eventually, clinics were held for haemophiliacs combining orthopaedic and dental care. The factor concentrates initially used in the treatment of haemophiliacs were prepared by the Blood Transfusion Centre in Dublin, but these were comparatively crude and were replaced by more refined concentrates from the USA which were capable of being used for home treatment. In the early 1980s these were found to be contaminated with the Aids virus, which tragically led to the death of a number of haemophiliacs in Ireland, as in many other countries.

In 1969 Dr Edmund Bourke was appointed by Trinity College and the Federated Hospitals as lecturer in medicine and nephrologist and was assigned to the Meath Hospital. He had had an outstanding academic career and had been trained in the Westminster Hospital, London, and the Georgetown University Hospital, Washington. This appointment met the needs of the GU unit in providing advice and treatment for medical renal cases. Dr Bourke stimulated considerable research in association with the biochemistry department at Trinity College, which was supported by the Wellcome Trust and other research foundations.

In 1971 Dermot O'Flynn founded the Irish Stone Foundation, under the chairmanship of R M Graham, for research into the cause and treatment of renal stones, and initially Edmund Bourke and his research assistants were closely associated with this work. By 1980 up to £80,000 had been raised by volunteers for the foundation and, after this, financial support was extended to all aspects of kidney disease. Frank Delaney, Dr Ann McInerney, Lottie Masterson and Barry McHugh all gave major assistance in raising funds over the years.

The first lithotripter in Ireland, for the non-operative treatment of renal stones, was acquired by the GU unit in 1987. This was accomplished by the energy and enterprise of Michael Butler and Gerard Hurley. This equipment cost £700,000: a capital grant of £400,000 was obtained from the Department of Health and the balance was collected by a fund-raising campaign. Patients were referred for this treatment from all over Ireland.

Although Eddie Bourke was a brilliant teacher and tireless in his attention to patients suffering from complicated biochemical problems, he could be unconventional at times. In 1974 he ordered four haemodialysis machines for the Meath without going through the normal administrative channels, the large packing cases arriving unannounced in the west wing corridor. There was considerable protest but, nevertheless, the equipment was installed in the wooden extension at the back of the west wing which was known as St Mary's and was originally built for TB cases. Until then the unit at Jervis Street Hospital provided the only haemodialysis facility in Dublin, so there was undoubtedly a need for further treatment facilities. There had always been close co-operation between the Meath and Jervis Street Hospital. The first transplant took place in 1977 and, by 1985, 69 patients from the Meath had had successful transplants. Over three hundred renal transplants had been carried out on Meath Hospital patients by the end of 1995, including two combined kidney/pancreas transplants. There were approximately 165 successful renal transplant patients attending the transplant out-patient clinic in 1995 and, in the same year, approximately 7,500 haemodialysis sessions were carried out.

Since those early days, the Meath renal unit has developed under Brian Keogh as the main dialysis centre for the south side of Dublin. In 1979, to complement haemodialysis at the Meath and to expand the renal service, Keogh established the first CAPD (continuous ambulatory peritoneal dialysis) programme in these islands, widely accepted, both in Europe and the USA, as a complementary treatment to haemodialysis. This programme treats, on average, fifty patients per annum. There was an obvious need for modern accommodation for the renal unit at the Meath, and this culminated in the erection of a new building at the back of the west wing, which was opened by the Minister for Health, Dr R O'Hanlon, in April 1987. The building cost in the region of £500,000. A grant of £300,000 was given by the Department of Health and the rest was contributed privately. Mr Louis Cohen made a major donation towards the cost of the new building in memory of his brother, Israel, who died in the hospital in 1977.

In July 1972 Michael Cullen was appointed to a similar post to that of Bourke, but as a specialist in endocrinology. He had worked for several years in the Thorndike Laboratory, Boston, in thyroid metabolism, and he developed facilities for thyroid investigation at the Meath.

Specialised medicine was developing rapidly in the western world, and general physicians were inevitably being replaced by specialists. At the Meath, Brendan O'Brien and Brian Mayne were still in the old tradition of general physicians, although both had special expertise: O'Brien in chest disease, and Mayne in cardiology. My allotment of 25 beds had inevitably become invaded by specialised cases, under the care of Drs Temperley, Bourke and Cullen. The

Federated Hospitals' haematology department, centred at the Meath under the direction of Ian Temperley, now began to attract leukaemia cases, and this put considerable pressure on the medical bed accommodation. In 1977 the haemophilia unit (adult section) transferred to St James's Hospital.

Throughout this period, the Meath's X-ray department was expanding to meet increasing demands. Ultrasound and nuclear medicine units were developed there in 1977, under the direction of Sholto Douglas and greatly helped by Gerard Hurley, who was appointed radiologist in 1975. P V Doyle, the hotelier and member of the joint committee, gave valuable practical advice in building additional space for the department. Dr David McInerney was appointed as radiologist to the Meath and the Adelaide in 1979. Sholto Douglas, who was universally liked by his colleagues for his intelligent advice and sound opinion, retired in 1981. Under his direction the X-ray department had coped with excessive pressure, especially when the GU unit first opened. Computerised tomography (CT) was introduced in 1988 by Gerard Hurley, with the assistance of money raised by the Irish Stone Foundation.

In 1987 Dr Steevens' Hospital closed and its orthopaedic surgeons were transferred to the Adelaide and the Meath. This caused considerable disruption as regards bed allocation and operation sessions. John McElwain and David Fitzpatrick were assigned to the Meath and, together with Michael Pegum, were allotted forty beds for traumatic orthopaedics, thus reducing the number of beds for general surgery. After some negotiations, a floor of the genito-urinary unit was given over to general medicine, under pressure from the Department of Health, to meet the increasing number of emergency cases coming in from the busy accident and emergency department. Around this time gynaecological cases ceased to be accommodated at the Meath, and these patients were admitted to the Adelaide.

In 1990 Victor Lane retired from the GU department, having partnered Dermot O'Flynn over a period of more than 30 years. Both Dermot O'Flynn and Victor Lane were elected presidents of the College of Surgeons, thus bringing great distinction on themselves and to the Meath Hospital. Many surgeons to the Meath in the past had been elected to this office, the last being Douglas Montgomery in 1968. John Fitzpatrick was appointed GU surgeon in March 1981, but he only stayed for five years as he was appointed professor of surgery at University College Dublin in 1986 and thus transferred to the Mater Hospital. Ted McDermott and Ronald Grainger were appointed GU surgeons in 1987 and 1988 respectively and John Thornhill was appointed in 1990. Together with Michael Butler (appointed in 1974), this brought the number of GU surgeons to four.

Valuable Service

James Quin, gynaecologist, and Sydney Furlong, ear, nose and throat surgeon, retired in 1972. Both had given reliable and skilled service to the hospital, with good humour and kindness, for over thirty years. James Quin and his wife were active for many years in supporting the National Council for the Blind, especially in raising money to supply radios for the blind. Sydney Furlong, not unlike Lane Joynt, was 'good with his hands' and had a workshop at his home where he would make his own instruments, thus saving the hospital money. His colleagues would bring him all sorts of things which needed 'fixing'. He also flew his own aeroplane and used to land on the beach at Donabate; he presented an aero engine to Aer Lingus for their museum. Frank O'Loughran succeeded Sydney Furlong. He was ear, nose and throat surgeon at Dun's, a connection he retained when he was appointed to the Meath.

In 1973 Rory O'Hanlon was appointed to succeed James Quin, whom he had assisted for many years. Rory was a larger than life character who was commodore of the Royal St George Yacht Club and had crossed the Atlantic and circumnavigated Iceland in his yacht. He was compassionate and courageous in dealing with the problems of his patients. He was married to Dr Barbara Stokes, of the famous Stokes medical family, who herself was a distinguished paediatrician and campaigner for facilities for mentally handicapped children in Ireland. He died in 1979 and, according to his wishes, was buried at sea.

In November 1974 I resigned from the Meath, as I had been appointed medical director of the United Nations Medical Service in New York. I had served 17 years at the Meath, 22 years at Dr Steevens' Hospital and 14 years as professor of clinical medicine at Trinity. Eddie Bourke resigned in 1977 and went to the Ministry of Health and Medical School of Kuwait to assist in developing renal services in that country. Dr Brian Keogh was appointed in his place and the renal unit steadily expanded, rising from about 2,000 dialyses in 1979 to 3,782 in 1984. In 1978 Michael Cullen was appointed physician and endocrinologist to the Meath and St James's Hospitals. John Barragry, who had been consultant physician at King's College Hospital in London, was appointed physician in 1979 and, together with Brian Keogh, filled the vacancies created by the resignations of myself and Edmund Bourke.

Brandon Stephens retired in 1983, having spent 33 very busy years doing general surgery at the Meath. This included everything from the introduction of fractured femur pinning to the hospital to following Henry Stokes in parathyroid surgery. He also gave a great deal of time and energy to the hospital administration and was on the joint committee from 1960 to 1983. As chairman of the works committee he personally took an interest in renovating the west wing and installing the first public lift there. Pragmatic in nature, his favourite expression was, 'we must get the facts'. In his retirement he became an accomplished artist.

Brian Mayne officially retired in 1981 but continued on while awaiting his replacement, finally leaving in 1983, having served at the Meath for 36 years. Ian Graham took over the cardiac work of Brian Mayne in 1983 when he was appointed cardiologist to the Meath and the Adelaide. Colm O'Morain became gastroenterologist to the Meath and the Adelaide in 1984, and he was to found the endoscopy clinic.

In 1985 the secretary-manager, John Colfer, retired after 28 years of service. He had remained calm and patient through all the many daily administrative problems, which included dealing with infallible consultants or irate porters, maintaining good relations with the Department of Health and managing the expanding services that had to be provided by the hospital in old buildings on an overcrowded site. Dermot O'Flynn retired from his post of genito-urinary surgeon but stayed on until 1987 when his successor, T E D McDermott, was appointed. He had given a total of 35 years of outstanding service to the hospital.

CHAPTER EIGHTEEN

Nursing, Physiotherapy and Social Work

In 1807 only two nurses were employed in the hospital. This was increased to four in 1825 and seven in 1873. These nurses, who were often illiterate, did not receive any special training and were little more than domestic workers. Alcoholism was a problem at that time among the nurses, as they had control of the supplies of wine, spirits and porter ordered for the patients. Bessy Henry, who was found drunk on night duty, was discharged and 'her wages were to be paid up to her discharge, but her gratuity of one guinea was to be withheld'. Nurses then were poorly paid and remained employed until they were very old. On 30 December 1867 the standing committee noted:

> that Nurse Thompson should cease duty on account of her age of 75, and in view of her continuous and valuable services that it be recommended to the Lord Commissioners of Her Majesty's Treasury that she be granted a special pension at the rate of seven shillings per week.[111]

Organised disciplined nursing in Ireland was first provided by the Sisters of Charity and Sisters of Mercy in the 1830s. Instruction for non-religious was not available until the late 1850s, when most hospitals began admitting nursing probationers.[112] Initially most hospitals received probationer nurses from nursing institutions separate from the hospital. In 1859 the Adelaide Hospital opened its own training school for Protestant nurses.[113] A training institution for nurses for Dr Steevens' Hospital was founded at 152 James's

Street in 1866.[114] The Royal City of Dublin Hospital was associated with the City of Dublin Nursing Institution at 27 Upper Baggot Street which was established in 1883.[115]

The medical board of the Meath expressed dissatisfaction with the nursing in the hospital in 1879. Lambert Ormsby felt that the matron, Mrs E Jones, was not sufficiently trained, and so the governors arranged for her to attend the Royal Southern Hospital in Liverpool for some months to gain experience. However, the medical board judged that this was not adequate and eventually the situation came to the unfavourable attention of the committee of the Hospital Sunday Fund.

The Hospital Sunday Fund was formed when Charitable Sermons began to fail as a source of income for hospitals. From its earliest days, regular annual sermons had been held in aid of the Meath. A governor would be assigned to find an eloquent preacher who would be willing to give a fund-raising sermon in a neighbouring church, such as St Werburgh's. Prominent titled citizens would be asked to be collectors and the Lord Lieutenant and other notables would be invited to attend. About the middle of the nineteenth century these charity sermons became increasingly difficult to arrange, and the amounts collected eventually proved insignificant, so the Hospital Sunday Fund was set up to replace them and to collect funds centrally in aid of the Dublin hospitals on a designated Sunday. The money was distributed to the individual hospitals by a committee, and this committee seems to have become a sort of watchdog, whose view of the running of a hospital could affect the amount of money it received.

In 1881 the Hospital Sunday Fund asked the Meath to report on its training of nurses which had been questioned. In 1879 the medical board had recommended that Mrs Jones be relieved of her duties and a trained lady superintendent be appointed in her place, but the majority of the standing committee appeared to favour Mrs Jones, as she was very successful in obtaining donations for the hospital and the convalescent home, and so this advice was not followed. Judging by the comments of Lambert Ormsby and Philip Smyly, Mrs Jones did not provide nurses of good quality and she seems to have been more of a housekeeper than a nursing expert. In fact, in February 1883 she produced a comprehensive table of the stock of blankets in the hospital which is of some interest, as it included the exact distribution of beds in the hospital before the addition of the Barber–Bury Wing.

Accident (First Floor)	28
Surgical (Middle Floor)	31
Chronic (Medical, Top Floor)	38

Smyly (Children)	11
Fever (Epidemic Wards, Separate in Grounds)	13
Officials	9
Servants	2

Total 132 beds 171 pairs of blankets

The medical beds on the top floor, referred to as chronic, were the beds which had been restricted to fever cases in the past before the epidemic wards were constructed in the grounds.

Eventually the governors yielded and George Walpole, a lay governor, consulted with other hospitals and proposed that a training institute for nurses should be established, external and entirely separate from the Meath. Mrs Jones was persuaded to resign, being promised a pension of £40 per year for her nine years of service. The job was advertised for 'a fully trained Lady Superintendent who shall be competent to train nurses and supervise their entire duties'. There were eight applicants and Miss Ellinor Lyons was elected in February 1884, marking a turning-point for nursing at the Meath.

In October 1884 nurse probationers were employed from a nurses' home in Charlemont Mall, and by 1887 nurse probationers were employed from the Red Cross nursing home at 87 Harcourt Street. £40 per annum was paid by the hospital for every two nurses sent for training, and they were also given a dinner each day. By 1889 fifteen were taken every year for one year's training. The Harcourt Street home was under the supervision of the lady superintendent, Ellinor Lyons, and the training given in the hospital was directed by her and the staff sisters of the landings.

In July 1885 there was a viceregal visit to the Meath Hospital. The *Irish Times* report on 22 July 1885 of this rather grand visit contains a description of the hospital as seen by the public:

> At noon yesterday her Excellency the Countess of Carnarvon, accompanied by Lady Camilia Wallop, and attended by Mr Esme Howard, Private Secretary, and by Lieutenant the Hon L Ashburne paid a visit to the Meath Hospital and County Dublin Infirmary. Her Excellency was received at the entrance by Dr J W Moore, one of the physicians to the hospital, and the following members of the surgical staff—Sir George H Porter, Surgeon in Ordinary to the Queen in Ireland; Mr James H Wharton, Dr Philip Crampton Smyly, Mr Rawdon Macnamara, Mr Lambert Hempenstal Ormsby, Mr W J Hepburn, Mr F T Porter Newell (Resident Surgeon); and also by Miss Ellinor Lyons (Lady Superintendent).

Her Excellency brought two large baskets of flowers, which were carried round the hospital by the porters, and as she passed through the different wards, she gave a flower to each patient, accompanied by a few words of kindly sympathy. It should be remembered that each landing of the building is devoted to a separate department of hospital work, and that it was only the non-infectious portions that were seen. First the Accident wards, in which there are at present 25 cases under treatment, and the Smyly Children's ward, devoted to accidents and cases of surgical disease, which at present has a dozen cases, were visited. Next to the surgical landing, and from that the party went to the medical landing, and also to the medical children's ward. The operating theatre, the kitchen, and the storeroom were all in turn examined. After nearly an hour thus occupied, her Excellency went to the Boardroom, and signed her name in the visitors' book, at the same time taking occasion to state that she was greatly pleased with all she had seen and with the cleanliness and good order of the institution, which reflected much credit upon the Lady Superintendent. It should be mentioned that there are at present about 100 patients in all the institution, and that the average number relieved ranges between 90 and that figure; that in each department there is a lady probationer of the Red Cross Nursing Order, and that Miss Lyons is the head of the branch of that order in Dublin. Her Excellency took her departure shortly after one o'clock.

In 1900 there were 145 beds in use in the hospital, 79 medical and 66 surgical. The nursing staff, apart from Miss Lyons, consisted of five sisters, one night nurse, one surgical nurse, six two-year nurses and twelve probationers. The nursing costs were as follows: salary of the lady superintendent was £100 per annum; the sisters each received £20 to £30 per annum, plus £3 for uniform and a Christmas gratuity of one to two guineas; the six two-year nurses received no salary but were given £18 for uniform and six guineas Christmas gratuity; the twelve probationers were not paid anything. The average cost of food for each resident sister or nurse was £20 per annum, and for each non-resident (probationer) was £15 per annum.[116]

In 1904 the hospital considered taking over the Red Cross nursing institution in Harcourt Street but, for financial reasons, they decided against it. In 1904 a nursing committee of lay and medical members was formed, with Dr James Craig in the chair, and they decided that the four top wards in the west wing be allocated for accommodation of the nursing probationers. However, a nurses' home was eventually built in the grounds, to the east of the main building and Smyly ward, and connected to the main building at first floor level. This building was completed in 1907 at a cost of £4,700.

In 1906, the Red Cross badges which the nurses had worn until then were replaced by a metallic gilt and enamelled brooch bearing the arms and motto of the Meath Hospital. A pension scheme was also set up in 1906, to which the hospital and individual nurses were to subscribe. It was planned that every sister should retire, at the age of 55, on a pension of £30 per annum. Around this time the nursing committee received complaints that certain members of the medical staff were not giving lectures to the nurses as agreed. A rather pompous comment was received from the medical board that 'they consented to give such lectures as a voluntary act of courtesy and grace and therefore cannot subordinate their professional duties to the fulfilment of their undertaking'.[117]

With the outbreak of war in 1914, many nurses were recruited for war service. In 1916 a letter was received from the War Office requesting probationers for service for six months at a time for Voluntary Aid Detachments (VADs). A reply was sent that one or possibly two would be sent every three months and Queen Alexandra subsequently wrote thanking the hospital for answering the appeal.

In April 1919 the following resolutions were adopted at a meeting of representatives of the Dublin hospitals, in an effort to achieve uniformity:

1. That it is inadvisable to abolish probationers' fees.
2. That probationers should not receive payment during their three year training.
3. That hours of work be fixed at 56 hours per week.
4. That after three years training, a qualifying exam be held and certificates issued.
5. That four staff nurses fully trained and certified be appointed for one year and be paid a salary of £35 per annum.
6. That a night superintendent similarly qualified be appointed each year at £40 per annum.
7. That Staff Sisters be paid a minimum salary of £45 per annum.[118]

In June 1920 the salary for a sister was increased to £60 per annum and the uniform allowance was increased from £4 to £10. In the November, Lucy Dimond, who had trained in Sir Patrick Dun's Hospital, was appointed night sister, and a year later she became sister to the medical landing, where she gave distinguished service for thirty years. Dr Cyril Murphy said that her nursing skill was responsible for saving the lives of many patients suffering from lobar pneumonia before the introduction of antibiotics. Also in 1920, Sister Gage was put in charge of the west wing.

Matrons

After 24 years of service, the lady superintendent, Ellinor Lyons, resigned in October 1907 after a severe attack of typhoid. She was awarded a pension of 50 guineas per annum. Miss Laura Bradburne, known as Sister Florence when she had been in charge of the surgical landing, was elected in her place and the title of the post reverted to 'matron'. In August 1920, Laura Bradburne offered her resignation, but this was deferred for a year. She had been matron for 13 years and had entered the hospital as a probationer in 1889.

In 1922 Miss Mary C Wall was appointed matron and, shortly after her appointment, she recommended that the Meath's nurses should attend the Dublin Metropolitan Technical School for Nurses. The charge was £1 per nurse per course. This school was founded in 1893 by Margaret Huxley of the Eye and Ear Hospital and offered lectures in anatomy, physiology, hygiene and invalid cookery. By 1922 the curriculum was extended to include medical nursing, surgical nursing and gynaecology, and, later, lectures in medicine and surgery were added.

The General Nursing Council of Ireland was established in 1923, following the Nurses Registration Act (Ireland) of 1919. The Dublin Metropolitan School extended its curriculum to meet the new regulations in the training, examination and registration of nurses. Miss Wall advised that a tutor sister should be appointed to augment the teaching received at the Metropolitan School. The job would also include taking on the matron's duties during holidays. Sister Margaret Lytle was appointed to this post at a salary of £90 per annum. In August of the same year she complained to the nursing committee about the response of Sister Gage when she had asked her for a report on her patients. Miss Gage was interviewed and admitted having said to the matron that she was 'utterly contemptible'. The committee resolved that it would be 'impossible for these two officials to work together', but apparently peace was subsequently restored.[119]

By the 1930s the only hospitals in Dublin not affiliated with the Dublin Metropolitan School were Jervis Street, the Mater and St Vincent's, which were run by religious orders. Owing to the large number of students attending, the lectures were moved from 34 St Stephen's Green to the College of Surgeons. The class of approximately 400 had to be divided in two, to allow one half to attend to hospital duties and the other half to be accommodated in the lecture theatre. A lecture was given at 5 pm and was repeated the next day for the other half of the class. This was a tiring process for the lecturer and, inevitably, the lecture never had the same impact on the second day. In 1925 Nurse Eileen Garland of the Meath Hospital was awarded the gold medal by the school.

In 1928 Miss Wall retired to get married, after six years of service as matron. Sister Winifred Gage, who was in charge of the west wing, was elected in her place. The new matron, who was English and had been trained in London hospitals, was a tall striking figure and wore a cream-coloured veil and a brown dress. When the salaries of the nursing staff were reviewed in 1931, it was found that the salaries at the Meath were considerably lower than at the Adelaide, the Richmond, Dun's and the Royal City of Dublin. Consequently, the salary of the matron was increased to £200 rising to £250, that of the tutor sister was increased to £110 and the sisters to between £75 and £100.

In 1937 the nursing committee discussed the need to increase the period of training for nurses from three to four years, 'in order to fall into line with all other hospitals who have adopted the four years training owing to the difficulty in covering the syllabus required by the General Nursing Council'. The committee also decided that probationers should be paid during their training, and the annual salary was agreed as follows: 1st year—£10, 2nd year—£12, 3rd year—£14 and 4th year—£26.[120]

In 1943 proposals were made to improve the working conditions for senior personnel. This would include two whole days off per month and 'telephone leave' for sisters and nurses with over three years' service. The latter could only be availed of during a 'long' day on duty and only after 6.00 pm. Salaries of senior sisters were also raised from £120 to £150 per annum. Also in this year, Sister Eileen Sheridan, who pioneered for many years in the X-ray department under T J D Lane's direction, was confirmed as permanent X-ray sister and gave excellent service up to her retirement.

Though Ireland was neutral during the Second World War, many Irish nurses served in British hospitals or with the armed services. In December 1944, the hospital ship *Amsterdam* was torpedoed off the Normandy coast, but the patients were successfully evacuated owing to the bravery and calmness of the medical and nursing staff. Amongst these was Ellen T Hourigan of Limerick, a Meath hospital nurse and member of the Queen Alexandra's Nursing Service, who was awarded the MBE. In mid 1943 the matron, Miss Gage, applied for leave of absence to give service during the war and in the October Margaret Lytle, the assistant matron and tutor sister, resigned, so it was decided to appoint an assistant matron in her place to be acting matron till the return of Miss Gage. Miss Ann Magee was elected to the post. She had trained at Sir Patrick Dun's Hospital, where she had been theatre sister, and was a sister at the private clinic, Portobello House. Miss Gage returned to the position of matron in 1945, possibly with a fresh view of things after her sojourn abroad. On 26 May 1945, she expressed her opinion to Mr Lane via a letter from the secretary-manager, Mr Fitzpatrick:

> that it would be better from the point of view of the general running and administration of the hospital if specialist departments such as yours could be separately housed and staffed.

Mr Lane replied in writing to Mr Fitzpatrick on 5 July 1945:

> I am bound to admit that I consider the principle is sound. However, the practical difficulty of transfer, in my case at any rate, would be enormous and for that reason I should like to have any further details of any scheme projected.

Miss Gage retired in 1946 and Miss Ann Magee, the assistant matron, was promoted to matron in April. She had the task of trying to improve conditions for nursing in the hospital when facilities generally were rather run down after the Emergency years. She was concerned for the well-being of her nurses and was active in the planning of the new nurses' home which was opened in 1951. She was a strict disciplinarian and during her tenure of office several nurses were suspended from duty for variable periods if they were inefficient in their duties or stayed out late without permission.

Miss Magee retired in 1968 and Miss Elizabeth O'Dwyer was appointed matron in her place. She was very energetic and enthusiastically took part in planning improved conditions in the west wing, which was receiving the attention of the works committee. The west wing was an architectural nightmare at that time, having a double staircase and no lift. She was always considerate of the nurses' working conditions, and was instrumental in getting sisters' offices constructed in the old wards.

From the early 1980s there was increasing pressure on beds because of the expanding accident and emergency service. The unexpected closure of Dr Steevens' Hospital in 1987, and the transfer from there to the Meath of some of the orthopaedic service and nursing staff, caused additional strain on the running of the hospital, but Miss O'Dwyer dealt with these problems very successfully. She was always interested in the welfare of her nurses and assisted them discreetly when consulted about their personal problems. When it became acceptable for nurses to live outside the hospital, she supported this radical change, even though she could have opposed it in view of the capacity of the relatively modern nurses' home. She actively supported any means of improving the education of the nurses, promoting facilities for the on-going education for registered nurses and courses in specialised nursing. She also introduced the idea of a degree for nursing, which led to discussions between the MANCH group and Trinity College.

Other Nursing Personalities
Elizabeth Cunningham was the first nurse to be specially trained in urological nursing. She led a group of specialised nurses, which were essential for the new GU department which opened in 1955. As sister in charge of the department, she held the rank equivalent of assistant matron. She implemented the high standards demanded by Tom Lane in urological nursing and helped to make his unit famous. She died tragically in a car accident in November 1975.

The Hoey twins will be remembered by generations of nurses who trained at the Meath. Angela and Elizabeth Hoey entered as student nurses in 1953. They qualified in 1957 and both were outstanding nurses. Angela Hoey was nominated by Matron Magee for the first course for tutor sisters which was inaugurated at University College Dublin in 1960. She completed the course in 1962 and was appointed as the first qualified tutor sister at the hospital in 1962, second to Miss Eileen Forde, the senior tutor at the time. Sadly, Miss Forde died at an early age in 1968 and Angela Hoey then became principal tutor, during the steady expansion of the nurses' training school, until her retirement in 1994. Elizabeth Hoey, having been matron of the Dental Hospital, returned to the Meath as assistant matron in 1969 until her retirement in 1994. Both had given a total of forty years' service in nursing.

There are many other sisters who will be remembered with affection and gratitude by their patients and by the hospital staff, but it is not possible to mention them all. However, I will attempt to mention a few. Nora Lyons had a remarkable diagnostic flair and was in charge of the accident ward and the old gynaecology and ENT operating theatre on the first floor. She was feared and yet respected and liked by her nurses. Evelyn Doherty and Mary Duffy were popular ward sisters in the west wing. Sister Doherty later became assistant matron and her sister 'Snowy' Doherty was sister in charge of the out-patient and accident department before leaving to be married in 1956. Maureen Fallon was an outstanding theatre superintendent in the GU department and Josephine Mullen was an exuberant and dedicated sister who was a great admirer of Dr Mayne. Mena Lambert, who, from 1951, was theatre sister for general surgery for thirty years was quiet, unassuming and very competent. Few knew of her remarkable and distinguished service in surgical nursing in France and Egypt during the Second World War. In the early 1950s Roslyn

'Pinky' Casey assisted Dr Cyril Murphy in his allergy work, removing potential sources of allergy from the Dimond ward, such as feathers, mats, horsehair, etc. Later she became a tireless worker in the fund-raising and social events at the hospital, for which she had the ideal personality, as it was very hard to say no to her! In 1982 she was elected president of the Past Nurses' Union, which holds a general meeting each year combined with a social event.

The first physiotherapist to the hospital was Eva Cherry, sister of Jack Cherry, the orthopaedic surgeon at Dr Steevens' Hospital. Her appointment in 1931 was as a part-time masseuse; there was another masseuse then, also part-time, Miss Moynan. Miss Cherry became full-time in 1935 at a salary of £80 per annum and was the single-handed physiotherapist to the hospital until her retirement in 1965. At first she had to work in an inadequate room off the entrance hall and later in an equally small room in the west wing, despite her requests for better conditions. She was a robust, cheerful personality, very successful in achieving early mobility in convalescent cases. Despite the lack of facilities, she gave remarkable service to the patients for 34 years.

In 1965 Miss Eileen 'Dan' Maloney was appointed physiotherapist to succeed Miss Cherry. She had qualified in Dublin in 1948 and, apart from her work in Dublin, had also worked in a hospital in California for two years. She fostered the development of the physiotherapy unit which she moved into a new Bantile building, initially sharing it as a lecture room and later expanding into the whole building, with a staff of assistants and students. She retired in 1990 and was succeeded by Helen O'Reilly, and then by Margaret Beswick in 1991. The present chief physiotherapist is Grace Cooke, who was appointed in 1993 and has eight qualified staff. The work of the department has greatly increased since the expansion of orthopaedics in the hospital in 1988.

The first lady almoner in the Republic of Ireland was appointed to the Adelaide Hospital in 1921. In 1926 the medical board of the Meath Hospital sent a resolution to the joint committee, proposed by Edward Lennon and seconded by Henry Stokes, strongly recommending the appointment of an almoner. No appointment was made, however, and it was not until the 1930s, following a report from the Hospitals Commission, that almoners were appointed to several Dublin hospitals. The Meath appointed its first almoner to the hospital in 1941. Miss Alma Brooke-Tyrrell remained for some years and was later succeeded by Mary Kane, who was assisted by Gretta Boylan. In 1949 Gretta Boylan was invited by the Sisters of Mercy to establish a social service department at the Mater Hospital, where she served for many years. Mary Lahiff was appointed in her place at the Meath.

At that time the Meath did not much value medico-social work. The joint committee decided to terminate the appointments of almoners Mary Kane and Mary Lahiff because their work was of 'no purpose', giving them notice that their employment would expire on 31 December 1956. Professor Jessop appealed against this decision, but Mary Kane left in December 1956. However, in March 1958, the committee confirmed Mary Lahiff's appointment as permanent and she continued on, by then known officially as 'medical social worker' rather than 'lady almoner', until 1990, assisted in succession by Heather Edge and Margaret Bradley.

In 1988 Michelle Hart, who had been medical social worker at Dr Steevens' Hospital, transferred to the Meath. She is assisted by three medical social workers, and their work is now fully appreciated by the medical staff. The department is allied with the social work departments of the Adelaide Hospital and the Children's Hospital, Harcourt Street.

The School of Nursing
The Meath Hospital School of Nursing reached a high level of efficiency during the 1980s. An innovative move at that time was to form a link with the Dublin Institute of Technology at Kevin Street to provide teaching in physics and chemistry and the social sciences as part of the regular teaching programme. The principal nurse tutor is Margaret McCarthy, who won the Nurse of the Year scholarship award in the education category in 1987 and now has ten teachers under her direction.

About fifty nurses qualify each year and the school also provides an in-service training programme for registered nurses and a post-registration operating theatre course. In 1952 Dr Cyril Murphy had founded a medal for the best nurse of the year, awarded for both clinical and theoretical ability, with money received from Georgina Wade. Two further prizes were instituted in memory of former nursing sisters of the hospital. The Lucy Dimond award is given to the student nurse who shows the most initiative and promise and the Nora Lyons award is presented to the registered nurse who makes the best contribution to the work of the hospital.

In 1993 the medical and nursing libraries were amalgamated and a professional librarian, Ms Ann Murphy, was appointed, which has encouraged the joint use of the teaching facilities by the medical and nursing staff. In recent years there has been increasing co-operation with the nursing schools of the Adelaide Hospital and the Children's Hospital, in preparation for amalgamation at the new hospital at Tallaght. In autumn 1996 the Meath nursing school, together with the schools of the Adelaide and of St James's, will commence a new diploma programme in nursing, in conjunction with the faculty of Health Sciences of Trinity College.

Chapter Nineteen

Social Occasions and Lighter Moments

The life of a hospital, like that of a human, does not consist of work alone. Social occasions for the medical, nursing and administrative staff and hospital supporters usually took the form of fund-raising events, particularly in the last century, when this was a typical feature of all voluntary hospitals, which were dependent on voluntary contributions for their very existence. In 1906 the Meath Hospital Guild was formed by a group of ladies for the purpose of providing clothes for destitute patients in the hospital. This was the forerunner of the 'ladies' committee' which replaced the hospital guild in July 1918. Then, in January 1944, the 'first meeting' of the present ladies' committee, chaired by Dr Angela Russell, took place. Ever since then, this committee has arranged many social events in order to raise money for items for patient care and comfort which are not routinely financed by the hospital.

The Smyly children's ward, founded in 1865, was supported by charity, independently of hospital funds, for about fifty years. Many events were organised to raise money for its maintenance during the Victorian period, the 'gypsy bazaar' in Bushy Park, Terenure, in 1886 being one such example. A grand ball was held in February 1906 at the Royal Dublin Society to raise money towards the pension fund for nurses. The honorary secretary to the 'ball committee', which raised £500 9s 0d, was surgeon William Taylor and:

> The Countess of Mayo induced their Excellencies the Lord Lieutenant and the Countess of Aberdeen to honour the ball with their presence.

In December 1920 the Dublin associated hospitals' fête was held and £2,000 of the proceeds was allotted to the Meath. In August 1921 a garden fête was held in the home of Mrs Collis at Kilmore, Killiney and, in January 1922 David Telford, hospital treasurer, announced that the proceeds of the Meath Hospital fête were £2,370 7s 1d.

As the hospital became increasingly supported by state funds, voluntary fundraising was aimed at providing specific accommodation, equipment, or research for which government money was not available. In recent years, these events were usually organised by the ladies' committee, chaired by Mrs Mary Mulvey and later by Mrs Kitty O'Brien, or the Irish Stone Foundation. The funds raised in this way enabled construction of the annexes to the hospital, St Catherine's for genito-urinary tuberculosis and St Mary's for pulmonary tuberculosis.

From early in the last century, especially during the famous days of clinical teaching at the Meath, meetings were held every October to open the 'clinical session', at which an inaugural address was given by a senior member of the medical staff or some other distinguished person. Many leading figures in the government and medical world attended these occasions, and the subject was often of medico-political importance. Robert Graves, William Stokes, and Oliver St J Gogarty were amongst those who spoke in the past and gave memorable addresses. The custom ceased from the time of the First World War but was revived in the 1940s. Widely reported in the press at the time, the following lectures were given, and published in the annual reports:

1943 Tom Lane, 'Notes on the Hospital Problem'
1944 Robert Steen, 'Looking Back Twenty Years'
1945 Henry Stokes, 'The Meath Hospital and County Dublin Infirmary'
1946 Brendan O'Brien, 'Tuberculosis and the Meath Hospital'
1947 Cyril Murphy, 'The Hippocratic Oath'
1948 James Quin, 'The Future of Medical Practice'
1951 Tom Lane, 'The Case for Specialisation'
1952 J D H Widdess, 'The Meath Hospital and County Dublin Infirmary 1753–1900'

This tradition for the opening of the clinical session has since ceased and the AGMs, usually held in April of each year, alone continue to give an impression of the hospital to the public. However, while all these occasions gave an opportunity for socialising between the governors, doctors, nurses and hospital supporters, they were formal occasions, unlike the annual hospital dances, which were an occasion for the younger members of the medical staff and the nurses to relax.

Meath Hospital Dinner

The most important annual safety valve for the medical staff and residents was the famous 'Meath Hospital dinner'. The first of these was held in 1883 in the Shelbourne Hotel, but there are few records of these occasions until 1926, since when it has become an annual event—except during the war years of 1940 to 1945—and records have been faithfully kept. Cecil Robinson organised this event until 1964, followed by his son Derek. The guests usually included distinguished members of other medical institutions. There was a special informal atmosphere about a Meath Hospital dinner which allowed for speeches to include comments on the peculiarities of the senior medical staff. Menus have been illustrated by appropriate cartoons and ballads have been sung to illustrate hospital idiosyncrasies. A silver loving cup, a gift from Henry Stokes, is filled with champagne and passed around for all to sip from.

Past students of the hospital were often invited to the dinner. In 1937 Dr W H Waterfield of Plymouth wrote regretting that he was unable to accept the invitation. In his letter he recalled an episode experienced by many other medical students on their first day.

> How well I remember Nov 1879 climbing up the hospital steps in fear and trepidation of what was going to happen, then joining a group of fellow students to go round the wards with Mr W J Hepburn. How long that lasted I never knew for I found myself lying on a couch in the residents' room being given a glass of port wine. It appeared I had fainted on seeing Mr W J H pass a catheter, and as I walked down Long Lane home I thought I had better tell my people that I would be more useful keeping white mice than learning medicine. But thank goodness that was my first and last faint and the words of Rawdon Macnamara came true, for the next day he said to me, 'Ah me boy don't worry, sure Napoleon fainted at his first battle and look what he did.' Another good advice this quaint surgeon gave was, 'Gentlemen, if at any time you are in doubt, believe a woman's belly before her Bible oath'—he was lecturing on pregnancy.[121]

Medical Student Misbehaviour

Medical students are renowned for their unruly behaviour off duty. The usual excuse is that young people who have to work very hard, and who are suddenly plunged into an atmosphere where suffering and death are commonplace, need some relief. The minutes over the last 200 years occasionally feature disciplinary measures taken for the bad behaviour of medical residents. Venturing into the nurses' home was not an unknown crime!

In 1937, with complete lack of respect for the famous of the past, the students removed the busts of Stokes, Collis and Crampton from the front hall and staged a mock-up party with them in the students' residence. Fortunately the busts were undamaged. The most spectacular prank occurred in 1958, when the senior resident's 1935 vintage car was placed in the front hall. It would have required considerable strength to move the car up the fifteen stone steps of the main entrance, and its removal to terra firma was photographed and recorded in the newspapers of the day.

Chapter Twenty

Some Reflections

The appointment of a relative of a member of the staff of a hospital to a post of physician or surgeon is common enough to this day. Although it can be argued that undue influence may have been exercised, this can be perfectly justifiable when the appointee has the required qualifications and competes in open competition. There is also a good chance of this happening because medicine tends to 'run in families'. In the case of the Meath Hospital, the incidence of family appointments was exceptionally high, particularly so because medical appointments were completely in the hands of the medical board, as laid down by the Acts of 1774 and 1815, and this privilege was jealously guarded. However, this did create a reputation for the Meath, right through the Victorian period and into this century, of inbreeding and nepotism.

Though influence may have played a part in these family appointments, in general they proved to be very successful. The various members of the Stokes, Smyly and Collis families served the Meath Hospital well and all had very distinguished careers. The elections of sons or nephews, with their dates of appointment, are as follows:

Richard Dease, surgeon (1795), son of William Dease, surgeon (1793–8).
Cusack Roney, surgeon (1802), and Thomas Roney, surgeon (1813), both sons of Patrick Cusack Roney, surgeon (1782–1813).
Josiah Smyly, surgeon (1804), nephew of Philip Crampton, surgeon (1798–1858).

William Stokes, physician (1826), son of Whitley Stokes, physician (1818–26).

Rawdon Macnamara II, surgeon (1861), son of Rawdon Macnamara I (1819–36).

George Hornidge Porter, surgeon (1849), son of William Henry Porter, surgeon (1819–61).

Maurice H Collis, surgeon (1851), nephew of Maurice Collis, surgeon (1825–52).

Philip Crampton Smyly, surgeon (1861), son of Josiah Smyly, surgeon (1831–64).

William Stokes, surgeon (1864), son of William Stokes, physician (1826–75).

Charles Ross, physician (1910), nephew of Edward Lennon, physician (1894–1940).

William Boxwell, physician (1911), nephew of William Stokes, surgeon (1864–8 and 1888–1900).

Henry Stokes, surgeon (1912), nephew of William Stokes, surgeon (as above)

Cyril Murphy, physician (1928), nephew of Edward Lennon, physician (1894–1940).

Medical Teaching

From the beginnings of the hospital in the late eighteenth century, when the hospital was dominated by surgeons, teaching was by the apprenticeship system, whereby the pupil was bound to the surgeon for five years. This system had died out by the 1880s. In the 1820s Robert Graves introduced bedside teaching and the assigning of students to patients whose cases they had to follow up and record. This approach was widely adopted by other schools both in Britain and North America. The teaching reputation of the Meath was at its height during the time that Graves and Stokes were physicians together, from 1822 to 1847.

The student fee book shows that there were 83 students registered in the year 1833. The fee for twelve months' instruction was 18 guineas and the total fees paid that year amounted to about £1,490. Some money was kept back by the medical board to pay for student prizes and other expenses and the remainder would have paid a dividend of about £200 each to the seven members of the medical board. This was a significant sum of money at the time and was the only remuneration received by the medical staff at the hospital. The dividend from teaching fees was recorded by the medical board over the years and this indirectly gives an idea of the volume of student attendance each year. In 1837 the total fees collected amounted to £651 and the dividend was £81 each. In 1872 the dividend was £89, and it rose to £115 in 1878, a 'record year'. The dividend in 1890 was £102, but thereafter it varied from £31 to £85. During the First World War it ranged from £48 to £60, but there was a post-war boom in the number of students between 1919 and 1922, with the dividend rising to £137. A similar rise in student numbers occurred after the Second World War, when the annual figure rose to between £83 and £195.

From the 1830s the medical board agreed on a teaching programme and they had a brochure printed. Each physician or surgeon was assigned a day for his bedside teaching. The class was usually held at 9.00 am and was known as a Clinic or Clinique. This arrangement for a senior physician or surgeon, now known as a consultant, to personally teach a class of medical students at the bedside became an accepted form of instruction. This method was peculiar to Irish medical schools, whereas in British schools the clinical teaching largely consisted of tutorials given by junior hospital doctors, of clinical clerkships and ward rounds during which students would receive only irregular attention.

Before 1953, when hospital internship became compulsory for all medical graduates, there were only four or five qualified resident medical officers in the hospital. For example, two house surgeons, two physicians and a casualty officer comprised the resident qualified staff at the Meath in 1945. The other residents, usually about six, were all students who did a great deal of the practical work both in the wards and in the accident department. Student residence in hospital, with considerable clinical responsibilities, was a distinctive form of Irish medical education up to the 1950s. As a result, newly qualified Irish doctors, even without having done a house job, were renowned for their practical expertise and initiative, in comparison with medical graduates from other medical schools.

From the middle of the nineteenth century, senior, intermediate and junior clinical examinations in medicine and surgery were held each year. Gold and silver medals were given for first and second prizes in the senior examination up to 1957. In 1924 a legacy of £600 was received from the late Dr Joseph Smyth of Naas to establish a prize. This gained in capital value over the years and the annual value of the prize, known as the Smyth Memorial Scholarship, varied between £50 and £100. In 1943 Mrs R A P Rogers, daughter of Sir John Moore, gave £75 to found a prize in medicine for junior students. Initially, the Sir John Moore Prize was given to the student with the highest marks in an examination on the use of the stethoscope and urine examination. Up to recent times these hospital examinations were arranged by the medical board. The prizes were competed for in the hospital setting and gave an opportunity for students with clinical expertise, rather than strictly scientific knowledge, to gain a distinction which would be useful in their future career.

CHAPTER TWENTY-ONE

Towards Tallaght

The Meath Hospital and County Dublin Infirmary is coming to the end of its long life of 244 years. In 1997 its energy and tradition will be absorbed into the new Tallaght Hospital, together with the Adelaide Hospital and the National Children's Hospital.

Until well into this century, the physicians and surgeons constituting the medical board dominated the style and tradition of the hospital by jealously exercising their unique legal right of electing their successors. This privilege favoured the trend for appointments to be made from medical families, the most notable of these being in the names of Stokes, Smyly and Collis. Through the Victorian era, gifted members of these families were among the leaders of the medical profession in Ireland.

Though the hospital was interdenominational in character, the Protestant influence persisted in the medical staffing of the hospital. Catholic activists believed that the situation at the Meath justified voting out the mixed hospital board and replacing it by purely Catholic members. This was achieved in 1949 but caused dissension and a bad atmosphere in the hospital. The majority opinion of the public at the time was that this was an extreme and unwarranted action. This led to a special Act of Parliament in 1951 which reconstituted the board of governors, giving it the legal authority to make the medical appointments instead of the medical board. Incidentally, in 1974 an interdenominational chapel was established in a quiet area in the modern section of the hospital and a service of dedication was held on 7 August, conducted by the Revd J Lawlor, O Carm., the Revd W Smallhorne, Church of Ireland, and the Revd E Gallagher, Methodist.

Conservative in nature and always limited by financial constraints, the governors and the medical staff were slow to move into specialisation and the development of ancillary services during the first half of the twentieth century. The exception to this was T J D Lane. By his foresight, determination and industry, he was responsible for developing one of the best specialised and modern surgical departments in these islands, thus dramatically advancing standards in specialised urological surgery by the 1950s. This development created a marked contrast between Lane's modern department and the remainder of the hospital, still housed in old buildings.

Stimulated by Lane's example and by the rapid advance in medical knowledge and expertise, from 1960 onwards there was a steady development in specialisation and expansion of services at the Meath, despite the very inadequate and out-of-date accommodation. This development was hindered by the fact that the Meath was one of the three hospitals on the south side of Dublin designated to take the continuous stream of accidents and emergency cases. Nevertheless, despite this, departments of renal medicine and orthopaedic surgery have been developed, special wards for coronary care, intensive care and day care have been long established and gastroenterology has a very active unit.

The nursing school has expanded and encouraged training in specialised nursing. The ancillary departments of physiotherapy, social work, dietetics and occupational therapy have all been further developed. The X-ray, or imaging, department performs up to 60,000 examinations a year, which includes ultrasound, isotope scanning and computerised tomography. To meet all these demands on a limited hospital site, numerous temporary buildings and ugly extensions have had to be built. This has resulted in marked congestion on the old Meath Hospital site, which in 1822 had its single original building surrounded by lawns and trees.

If there is a lesson to be learnt for the future from perusing this history, it surely must be that differences amongst the medical staff or governors are best resolved by themselves, internally, at committee level. Open and public dissension and hostility damage the reputation of the hospital, and the resulting tension within the hospital may ultimately interfere with the care of the patients.

A Bright Future
Discussions between the representatives of the Meath and Adelaide hospitals about the building of the hospital at Tallaght have continued intermittently

since 1977. In 1981 the government confirmed the Establishment Order for the Tallaght Hospital and altered it to include the National Children's Hospital. By 1984 the Meath, Adelaide and Children's Group was generally known by the acronym MANCH, and the central council of the Federated Hospitals formed a Manch Council.

There were protracted discussions between the representatives of the three hospitals in the 1980s about the future administrative structure of the amalgamated hospital. In 1990 the government appointed Mr David Kingston, chief executive of Irish Life, and Mr David Kennedy, former chief executive of Aer Lingus, to chair a committee of hospital representatives of the Manch group in order to reach agreement on the administration of the Tallaght Hospital. In 1992 the following agreements were reached about the new hospital:

- It will be called the Adelaide and Meath Hospital.
- It will be a public voluntary teaching hospital and be multi-denominational and pluralist in character.
- The Church of Ireland Archbishop of Dublin will be the president of the hospital.
- The Adelaide and Meath Hospitals will each appoint six representatives to the new board and the National Children's Hospital three. Each hospital will include one senior medical representative.
- The Minister of Health will appoint eight representatives, six being nominated by the president of the hospital.
- There will be one representative from the Eastern Health Board and one from the medical school.

The text of the Charter of the Adelaide and Meath Hospital, Dublin, incorporating the National Children's Hospital, was agreed by the chairmen of the three hospitals on 28 April 1995. The text is fully consistent with the letter and spirit of the 'Heads of Agreement' approved by the boards of the Adelaide, Meath and National Children's Hospitals in May 1993 and endorsed by the government.

The staff now look forward to moving to modern accommodation in Tallaght, with appropriate dedicated space for them to carry on their excellent work. Co-operation has taken place at all levels and in all departments and is motivated by the determination of all the staff, both of the Meath and its sister hospitals, the Adelaide and the Children's, to make the new Tallaght Hospital a success by continuing their great tradition in providing a first-class service to the sick.

NOTES

*Min = Minute Book; SC = Standing Committee;
JC = Joint Committee; MB = Medical Board.*

1. J D H Widdess, 'The Meath Hospital and County Dublin Infirmary' in *Annual Report and Bicentenary Yearbook 1753–1900* (Dublin, Brunswick Press 1953), p 21.
2. C Maxwell, *Dublin Under the Georges* (Faber and Faber 1937), p 139.
3. *Ibid*, p 140.
4. *Ibid.*
5. W Boxwell, 'Historical Sketch of the Meath Hospital and County Dublin Infirmary' in *The Medical Press and Circular*, April 1 and 8 (1942), vol CCVII.
6. Deirdre Lindsay, *Dublin's Oldest Charity* (Anniversary Press 1990).
7. Widdess, *op. cit.*
8. Min SC, 6 April 1807.
9. Frederick Mullally, *The Silver Salver: The Story of the Guinness Family* (Granada 1981).
10. L H Ormsby, *Medical History of the Meath Hospital and County Dublin Infirmary* (Dublin, Fannin; London, Bailliere, Tindall and Cox 1888).
11. Min SC, 20 April 1807.
12. Min SC, 17 April 1809.
13. Davis Coakley, 'John Cheyne, 1777–1836' in *Irish Masters of Medicine* (Dublin, Town House 1992), p 65.
14. Min SC, 12 April 1813.
15. Widdess, *op. cit.*
16. Mullally, *op. cit.*
17. B Bayley-Butler, 'Thomas Pleasants' in *Dublin Historical Record* (1944), vol VI, p 121–32.
18. Ormsby, *op. cit.*, p 115.
19. Min MB, 31 July 1821.
20. Ormsby, *op. cit.*, p 22.
21. Ormsby, *op. cit.*, p 46.

22. W Gibson, *Rambles in Europe in 1839* (Philadelphia, Lea and Blanchard 1841).
23. R Graves, *London Medical and Surgical Journal*, VII, 2 (1835), p 516.
24. W Stokes, *Diseases of the Chest* (Dublin, Hodges and Smith 1837).
25. W Stokes, *Diseases of the Heart and Aorta* (Dublin, Hodges and Smith 1854).
26. D Coakley, *Irish Masters of Medicine* (Dublin, Town House 1992).
27. Coakley, *op. cit.*, p 95.
28. *Medical Times and Gazette*, quoted from R Graves, *Clinical Lectures on the Practice of Medicine* (New Sydenham Society 1884).
29. J B Lyons, *Brief Lives of Irish Doctors* (Blackwater 1978), p 63; S Taylor, *Robert Graves* (London, Royal Society of Medicine 1989); Coakley, *op. cit.*, p 89; D Coakley, *Robert Graves, Evangelist of Clinical Medicine* (Irish Endocrine Society 1996).
30. W Stokes, *William Stokes* (London 1898); J B Lyons, *Proceedings of the xxiii Congress of the History of Medicine* (London, September 1972), p 1010; Lyons, *Brief Lives, op. cit.*, p 84; Coakley, *op. cit.*, p 123.
31. Widdess, *op. cit.*
32. Gibson, *op. cit.*
33. Ormsby, *op. cit.*
34. Martin Fallon, *Sketches of Erinensis* (London, Skilton and Shaw), p 100.
35. Min MB, 1824.
36. W McN Dixon, *Trinity College Dublin* (London, Robinson and Co 1902), p 175.
37. Min SC, 8 June 1826.
38. R Graves, *Clinical Lectures on the Practice of Medicine*, vol I (New Sydenham Society 1884), p 97, 98.
39. F H Garrison, *History of Medicine*, 4th edn (W B Saunders 1929), p 706.
40. Graves, *op. cit.*, vol I, p 107.
41. Graves, *op. cit.*, vol I, p 191.
42. E O'Brien, *Conscience and Conflict: A Biography of Sir Dominic Corrigan, 1802–80* (Dublin, Glendale 1983), p 220–1.
43. Maxwell, *op. cit.*, p 147.
44. O'Brien, *op. cit.*
45. Min SC, 21 August 1826.
46. C Woodham-Smith, *The Great Hunger* (London, Hamish Hamilton 1962), p 188.
47. Min SC, 29 Jan 1849.
48. Woodham-Smith, *op. cit.*
49. Graves, *op. cit.*, vol I, p 123.
50. Ormsby, op. cit., p 137.
51. D Mitchell, 'A peculiar place' in *The Adelaide Hospital, Dublin, 1839–1989* (Blackwater 1989), p 54.
52. Coakley, *op. cit.*, p 321.
53. Coakley, *op. cit.*, p 99.
54. Lyons, *Brief Lives, op. cit.*, p 86.
55. *Irish Times*, 1, 8, 11, 15, 21, 31 July 1861.
56. *Ibid.*
57. *Ibid.*
58. *Ibid.*

59. *Ibid.*
60. Letter in hospital records.
61. Lyons, *Brief Lives, op. cit.*, p 88.
62. Ormsby, *op. cit.*, p 214.
63. Ormsby, *op. cit.*, p 222.
64. Widdess, *op. cit.*, p 40.
65. Ormsby, *op. cit.*, p 218.
66. Ormsby, *op. cit.*, p 226.
67. Ormsby, *op. cit.*, p 230.
68. Min SC, January 1869.
69. Mitchell, *op. cit.* p 60.
70. Annual Report, Meath Hospital, 1873.
71. Ormsby, *op. cit.*
72. Annual Report, Meath Hospital, 31 Mar 1892, p 57.
73. Ormsby, *op. cit.*, p 239.
74. Min SC, 12 Mar 1894.
75. D Wormell, personal communication.
76. Boxwell, *op. cit.*, p 26.
77. R Kee, *The Green Flag* (London, Weidenfeld and Nicholson 1972), p 436.
78. Min MB, 9 June 1898.
79. *Irish Times*, 21 April 1900.
80. J B Lyons, *An Assembly of Irish Surgeons* (Glendale 1983), p 46.
81. Min JC, December 1911.
82. Mitchell, *op. cit.* pp 191, 331.
83. D G Browne and E V Tullett, *Bernard Spilsbury: His Life and Cases* (Harrap 1951), p 274.
84. Mitchell, *op. cit.*, p 263.
85. R Collis, *The Silver Fleece* (London, Thomas Nelson 1940) p 43.
86. R Barrington, *Health, Medicine and Politics in Ireland, 1900–70* (Institute of Public Administration 1987), p 108.
87. Coakley, *op. cit.*, p 321.
88. D McInerney, 'Development of hospital X-ray departments in Dublin' in J Carr (ed.) *A Century of Medical Radiation in Ireland* (Anniversary Press 1995), p 30.
89. Ormsby, *op. cit.*, p 233 (second edition).
90. Annual Report, Meath Hospital, 1873.
91. R Collis, *To Be a Pilgrim* (London, Secker and Warburg 1975).
92. J B Lyons, *Oliver St J Gogarty* (Blackwater 1980), p 282.
93. Lyons, *Brief Lives, op. cit.*, p 88.
94. Annual Report, Meath Hospital, 31 Dec 1943, p 47.
95. *Irish Times*, 12 April 1949.
96. *Irish Times*, 7 May 1949.
97. E Bolster, *The Knights of St Columbanus* (Gill and Macmillan 1979), p 109 and 111.
98. *Irish Times*, 3 October 1949.
99. *Irish Times*, 8 August 1949.
100. *Irish Times*, 30 July 1950.

101. 50th anniversary conference brochure, Irish Health Services Management Institute, November 1995, p 35.
102. *Irish Independent*, 17 November 1950.
103. Personal memoir by Professor Jessop.
104. *Ibid.*
105. *Irish Press*, 17 February 1951.
106. *Irish Times*, 1 March 1951.
107. *Irish Times*, 28 April 1951 and 1 May 1951.
108. Annual Report, Meath Hospital, 1951.
109. R Barrington, *op. cit.*, p 234.
110. D O'Flynn, *Journal of Irish Medical Association*, vol 60, 1967, p 105.
111. Min SC, 30 Dec 1867.
112. P Scanlan, 'The Irish nurse' in *A Study of Nursing in Ireland: History and Education 1718–1981* (Drumlin Publications 1991).
113. D Mitchell, *op. cit.* p 86.
114. T P C Kirkpatrick, *The History of Dr Steevens' Hospital* (Dublin University Press 1924), p 282.
115. D Coakley 'Baggot Street' in *Royal City of Dublin Hospital* (1995), p 29.
116. Min SC, Nov 1900.
117. Min Nursing Committee, 26 Feb 1909.
118. Min Nursing Committee, 12 Apr 1919.
119. Min Nursing Committee, 20 Aug 1923.
120. Min Nursing Committee, 10 Nov 1937.
121. Letter in Medical Board files.

APPENDIX I

Act of Parliament
George III, Regis, Cap LXXXI

An Act to amend several Acts for the management and direction of the Meath Hospital and County Dublin Infirmary, and for the better regulating the same. 14th June, 1815.

Clause VII

And it be further enacted, that the present Physicians and Surgeons of the said Meath Hospital or County of Dublin Infirmary, are hereby declared the Physicians and Surgeons of said Hospital or Infirmary; and that it may be lawful for said Physicians or Surgeons, or a majority of them, to elect a Physician or Surgeon in the room of any Physician or Surgeon who from time to time by death, removal, or otherwise, shall make a vacancy in the said Hospital; such Physicians and Surgeons nevertheless to continue to attend and serve at said Hospital or Infirmary without fee, salary, or reward: Provided always that all such elections shall be from the Members or Licentiates of the King's and Queen's College of Physicians, and Royal College of Surgeons in Ireland, and that a notification in writing of all such elections shall be duly made, by such Physicians or Surgeons, to the said Committee; and provided also that the said Committee shall have a power, and they are hereby authorized at all times, to suspend or dismiss any of such Physicians, Surgeon, or Surgeons, for neglect or improper conduct; and that no Physician or Surgeon so dismissed shall in future be eligible to be re-elected, unless he shall be previously recommended to the Physicians and Surgeons by the said Committee to elect and appoint a Physician or Surgeon in the room of any Physician or Surgeon who from time to time by death, removal, or otherwise, or by being dismissed for neglect of duty or improper conduct, shall make a vacancy in said Hospital, if the said Physicians or Surgeons shall decline or neglect to fill up, elect, or appoint to said vacancy within the period of three months after such vacancy shall occur.

Appendix II

Officers and Staff of the Meath Hospital and County Dublin Infirmary

Chairman of Board of Management
(known also as Standing or Joint Committee)

1922	William S Collis
1935	Mrs M Cosgrave
1939	R W O'Brien
1948	Major T W Kirkwood
1949	Dr Herbert Mackey
1951	Alderman John McCann, TD
1956	Cllr Mary Mulvey
1967	Cllr Pearse Morris
1971	Cllr Mark Clinton, TD
1973	Robert M Graham
1984	Harold Ellis
1987	Cllr Austin Groome
1990	Peter Houlihan
1993	Cllr Gerry Brady

Matrons or Lady Superintendents

1800	Mrs A Gower
1808	Mrs C Bolton
1811	Mrs Mary Maiben
1831	Mrs Sarah Walker
1862	Mrs C McDonnell
1874	Mrs Enright (resigned)
1874	Mrs E Jones
1884	Ellinor Lyons
1907	Laura Bradburne
1922	Mary C Wall
1928	Winifred Gage
1946	Anne Magee
1968	Elizabeth O'Dwyer

Registrars or Secretary-Managers

1796	E Connell
1798	J Brady
1806	Luke Wall
1807	J A Bolton
1811	J L Monnet
1813	W Flintner
1816	W Savage
1817	A Maiben
1820	W J Smith

1821	W McCullagh		1871–1892	Arthur Wynne Foot
1822	R Whaley		1875–1933	John W Moore
1823	W McCullagh		1892–1910	James Craig
1825	E Mathews		1893–1940	Edward E Lennon
1829	J Ellis		1910–1911	Charles H G Ross
1835	B W Clarke		1911–1943	William Boxwell
1837	R Shaw		1928–1957	Cyril J Murphy
1845	E B Stanley		1933–1947	Robert E Steen
1881	Francis Penrose		1940–1980	W J E Jessop
1919	Robert Dow		1943–1974	Brendan O'Brien
1942	John Fitzpatrick		1948–1983	Brian Mayne
1950	F D Murray		1957–1974	Peter Gatenby
1954	J B Griffin		1971–1977	Edmund Bourke
1957	John Colfer		1974–	Michael Cullen
1985	E J Thornhill		1978–	J A Brian Keogh
1988	Nicholas Jermyn		1979–	John Barragry
1993	Derek Dockery		1983–	Ian Graham
			1984–	Colm O'Morain

Physicians

1754	Thomas Brooke		1989–	Gerald Tomkin
1754	Francis Hutchinson		1991–	Raymond Murphy
1756–1770	William Patten		1992–	Joan Power
1760–1767	John Donaldson		1993–	Desmond O'Neill
1767–1786	John Charles Fleury			

Surgeons

1770–1781	Daniel Cooke		1753–1781	Alex Cunningham
1781–1785	Francis Hopkins		1753–	Redmond Boat
1785–1793	Thomas Evory		1753–	David McBride
1786–1788	Edmund Cullen		1753–1781	Henry Hawkshaw
1788–1893	Daniel Bryan		1754–1784	James Mills
1793–1806	Thomas Bell		1754–1756	Henry Mapletoft
1806–1809	Francis Barker		1755–1793	William Vance
1806–1818	Thomas Egan		1756–	Michael White
1809–1811	Geo F Todderick		1767–	Mr Linley
1811–1817	John Cheyne		1777–1790	Arthur Winter
1817–1821	Patrick Harkan		1776–1795	Israel Read
1818–1826	Whitley Stokes		1781–1802	George O'Brien
1821–1843	Robert James Graves		1782–1831	Patk Cusack Roney
1826–1875	William Stokes		1784–1787	James Scott
1843–1861	Cathcart Lees		1781–1809	Benjamin Wilson
1861–1871	Alfred Hudson		1790–1819	Solomon Richards

1793–1798	William Dease		1974–	Michael Butler
1795–1819	Richard Dease		1979–1980	Bernard Fallon
1798–1858	Philip Crampton		1981–1986	John M Fitzpatrick
1801–1849	Cusack Roney		1982–	Frank B Keane
1809–1831	Thomas Hewson		1983–	Arthur Tanner
1813–1825	Thomas Roney		1986–1989	John Coolican
1819–1836	Rawdon Macnamara		1988–	David Fitzpatrick
1819–1861	Wm Henry Porter		1988–	Ronald Grainger
1825–1852	Maurice Collis		1988–	John McElwain
1831–1864	Josiah Smyly		1989–	Martin Feeley
1836–1861	Francis Rynd		1990–	John Thornill
1849–1895	Geo Hornidge Porter			
1851–1869	Maurice H Collis		**Clinical Assistants**	
1858–1858	Thomas Ledwich		1886–1893	Edward E Lennon (and pathologist, 1892)
1858–1888	James H Wharton			
1861–1904	Philip Crampton Smyly		1886–1892	James Craig
1861–1893	Rawdon Macnamara II		1893–1904	Arthur H White (and pathologist)
1864–1868	William Stokes			
1868–1868	James H Stronge		1898–1900	Richard Lane Joynt
1869–1879	Robert P White		1898–1900	William Taylor
1869–1871	Robert St J Mayne		1904–1911	William Boxwell
1872–1923	Lambert Ormsby		1936–1981	John O'Leary (supt of OPD)
1879–1911	Wm J Hepburn		1946–1948	Brian Mayne
1888–1900	William Stokes		1953–1960	John Phelan (urological dept)
1895–1900	R Glasgow Patteson		1955–1967	Derek L Robinson
1900–1922	William Taylor			
1900–1928	Richard Lane Joynt		**Assistant Physicians**	
1904–1911	F Conway Dwyer		1912–1921	Mather Thompson
1911–1939	Oliver St J Gogarty		1922–1928	Cyril Murphy
1911–1912	William Pearson		1929–1968	Cecil Robinson
1912–1955	Henry Stokes		1930–1933	Robert E Steen
1922–1967	Thomas J D Lane		1932–1934	W R F Collis
1952–1974	Douglas Montgomery		1964–1967	Donald Weir
1955–1983	E Brandon Stephens			
1963–1987	J Dermot O'Flynn		**Assistant Surgeons**	
1963–1990	Victor Lane		1928–1939	Sydney Furlong
1967–1971	David Lane		1928–1935	P T Somerville-Large
1971–1989	Derek L Robinson		1951–1955	E Brandon Stephens
1972–	Michael Pegum		1957–1967	David Lane
1974–	Bill Beesley		1967–1972	Derek L Robinson

Assistant Surgeons to Urological Dept

1952–1963	Dermot J O'Flynn
1952–1963	Victor Lane

Anaesthetists

1909–1912	Edmund Glenny
1912–	Henry W Mason
1919–	A Merrin
1923–1925	James Devane
1925–1928	Cyril Murphy
1928–1929	Mrs N Quin
1929–1972	Silvia Deane Oliver
1940–1961	Maureen Murphy
1952–1974	C H Wilson
1952–1954	Katherine Bradley
1953–1959	John Cussen
1953–1954	Michael McGrath
1956–1984	Patricia Mary Delaney
1956–1968	Fergus Quilty
1957–1959	Fiona Acheson
1959–1986	Kevin O'Sullivan
1961–1993	Peter Morck

(Appointments since federation)

1961–1993	Hugh J Galvin
1969–1994	Una Callaghan
1974–1988	Brigid M J Brennan
1984–	Enda Shanahan
1984–	Frank Breheny (resigned)
1981–	Mary Stritch
1985–	Barbara Eagar
1985–	Declan Magee
1988–	Gerard Fitzpatrick
1990–	Katrina O'Sullivan

Consultant Accident Surgeons

1972–1989	Derek L Robinson
1989–	Geoffrey Keye

Reserve Surgeons

1945–1952	R F J Henry
1945–1952	Douglas Montgomery

Otorhinolaryngologists

1911–1939	Oliver St J Gogarty
1939–1972	Sydney Furlong
1972–	Frank O'Loughran

Gynaecologists

1924–1938	Paul Carton
1938–1972	James Quin
1972–1980	Roderick H O'Hanlon
1980–1983	Hugo McVey
1983–1983	James Clinch
1989–1991	Michael Turner
1991–	Patricia Crowley
1992–	John E Drumm

Assistant Gynaecologists

1903–1917	Frederick W Kidd
1917–1924	Paul Carton
1951–1972	Roderick H O'Hanlon

Radiologists

1897–1900	Richard Lane Joynt (as clinical assistant)
1900–1912	Richard Lane Joynt (as assistant surgeon)
1912–1924	Henry W Mason
1924–1943	T J D Lane (also surgeon)
1933–1943	Cyril Murphy (medical)
1943–1952	Harold Pringle
1952–1981	Sholto J Douglas
1956–1986	Joan M T MacCarthy
1975–	Gerard D Hurley
1979–	David McInerney
1983–	Samuel Hamilton
1985–1986	J Horgan
1987–	Eric Colhoun
1988–1994	Noel O'Connell
1988–1995	John Gately
1995–	Risteard O'Laoide

Ophthalmologists
1918–1946	Euphan Maxwell
1946–1969	T J Macdougald
1969–1983	D H Douglas
1983–	Paul Moriarty

Dermatologists
1940–1964	Augusta Young
1964–1976	David Mitchell
1976–	Marjorie Young

Alienists or Psychiatrists
1935–1941	Richard Leeper
1942–1946	Robert Taylor
1946–1983	Norman Moore

Dentists
1913–1925	William Ogilvy
1925–1928	W Alexander Roberts
1928–1943	Gerald Hyland
1943–	Gerald H Owens

Pathologists
1886–1893	Edward Lennon
1893–1904	Arthur H White
1904–1911	William Boxwell
1912–1922	R M Bronte
1922–1924	T J D Lane
1924–1943	William Boxwell
1943–1946	J McGrath
1946–1956	R A Q O'Meara
1951–1952	D J O'Kelly
1956–1966	Joan Mullaney

Biochemist
1930–1980	W J E Jessop

Haematologists
1973–1995	Ian J Temperley

Apothecaries
1864	B Anderson
1867	W H Digges
1868	G Hope
1871–1876	F W Burkitt
1877	Fisher
1881	Henry William Oulton
1882–1889	Frank Porter Newell
1890	Andrew McFarland
1891–1923	Tenison Lyons
1923–1955	James Galashan

Pharmacists
1956–1960	B R Bailey
1960–1994	M O'Connor

Masseuses
1913–1925	Miss Moynan
1931–1965	Eva Cherry

Chief Physiotherapists
1965–1990	Eileen Maloney
1990–1991	Helen O'Reilly
1991–1993	Mary Beswick
1993–	Grace Cooke

Almoners or Medical Social Workers
1941	Alma Brooke–Tyrrell
	Mary Kane
	Gretta Boylan
1949–1990	Mary Lahiff
1975	Heather Edge
1978	Margaret Bradley
1988–	Michelle Harte

INDEX

Page numbers for illustrations and captions are in italics.
Mc, St and Dr are listed as if spelled out.

accident and emergency work, 2, 9, 10, 12, 25, 42, 55, 56, 76, 108, 127, 133, 154, 167, 173, 177, 178, 187, 190
Acheson, Fiona, 156
Act of 1774, 6, 130, 185
Act of 1815, 16, 41, 86, 119, 120, 130, 185
Adelaide Hospital, 1, 2, 38, 43, 46, 49, 79, 86, 94, 99, 109, 122, 157, 167, 169, 170, 176, 179, 180, 189, 190
administrative offices, 25, 162
'admissibility' of patients, 9–10
Aids virus, 165
allergy work, 135, 179
amalgamation (Tallaght Hospital), 1, 2, 99–100, *152*, 157, 180, 189, 190–1
anaesthesia, 13, 37, 78, 102, 115
Andrews, Thomas, 23
Anglo-Irish Treaty, 97, 98
apothecary, 27, 35, 50, 55, 91–2
apothecaries store, 12
see also salaries
appointments procedure
controversies, 6
between joint committee and medical board (1910), 80–7
Smyly appointment (1861), 40–1
see also Irish Times
for medical assistants, 101
Atkinson, Rt Hon John, 76

Baggot Street Hospital, 79, 114, 115–6, 157
Band, David, 128
Barber–Bury wing, 52–3, *58*, 108, 116, 171
Barker, Francis, 7, 8, 10

Barnett, Tommy, 159
Barragry, John, *150*, 168
Barrett, Dr 'Jacky', 28
Benson, Charles, 42
Beswick, Margaret, 179
Bewley, Anne, 32
Bibby, Thomas, 92
Board of Guardians, 49, 50
Board of Health, 30, 31
boardroom of Meath Hospital, 53, 107, 110
Boat, Redmond, 4, 6
Boer War, 78
Bolster, Evelyn (Sister M Angela), 122–3
Bonaparte, Napoleon, 15, 16
Bons Secours, Glasnevin, 132
Botanic Gardens, 16
Bourke, Edmund, 165–6, 168
Bovell, James, 23
Bowden, Joseph, 127
Boxwell, William, 5, 81, 82, 83, 84, 87, 103, 106, 113, 116, *139*, *140*
death of, 114
Boydell, William, 99, 121
Boylan, Gretta, 179
Bradburne, Laura (Sister Florence), *70*, *72*, 94, 97, 175
Bradley, Margaret, 180
Breen, Alderman John, 120
Brock, Gabriel, 121, 131, 162
Bronte, Robert Mathew ('Max'), 87–8, 100
Brooke, Thomas, 6
Brooke-Tyrrell, Alma, 179
Browne, Noel, 128
Bryen, Gerald, 5
Butler, Michael, *150*, 165, 167

Cameron, Sir Charles, 90

Campbell, Senator S P, 120, 121
Carmichael School of Medicine, 43, 49
Carton, Paul, 94, 100, 103, 110, 122, *140*
Casey, Roslyn 'Pinky', 178–9
casualty department *see* accident and emergency work
Catholics
on joint committee, 121, 122–3, 125, 132
on medical staff, 94, 189
opposed to 1953 Health Act, 136
see also medical board
Charitable Infirmary, Inn's Quay, 4
Cherry, Eva, 179
Cherry, Jack, 179
Cheyne, John, 1, 8, 10, 11
Cases of Apoplexy and Lethargy with Observations on Comatose Diseases, 11
Cheyne–Stokes respiration, 11
Childers, Erskine, 161, 164
cholera, 49, 50
civil war, 97–8, 100
clinical research, 159, 164, 165
clinical session inaugural address, 115, 182
Colfer, John M, 157, 169
Colgan, Senator Michael, 120, 121, 123, 127, *151*
Colles, Abraham, 21, 38, 42
Collier, Robert J, 90
Collins, General, 98
Collis, Maurice, 20, 25, 38–9, 45, 184
Collis wards, 38–9, 45
Collis, Maurice Henry, 38, 42, 43, 45, 46, 48, 52, *67*

203

INDEX

Collis, W R F (Robert), 38, 95, 106–7
Collis, William S, 38, 85, 95, 99
conditions in hospital, 8–9, 12
   in fever huts, 31–2
Connolly James, 96
convalescent home, Bray, 55, 56, 76, 159, 171
Convent of the Sacred Heart, Mt Anville, 79
Cooke, Grace, 179
Coombe, the *see* Dublin
Coombe Lying-In Hospital, 94
Cork Street Fever Hospital, 8, 9–10, 16, 18, 29, 34, 52, 89, 154, 155
coronorary care, 51, 164, 169, 190
Corrigan, Dominic, 32
Cosgrave, W T, 120, 121, 124
Costello, John A, 128
Cowan, P C, 98
Craig, James, 55, *70*, 76, 80, 81, 82, *137*, *138*, 173
Crampton, Philip, 5, 20, 24, 25, 39, 46, 51, *63*, 184
   teaching of, 24
Cross, Raymond, 126–7
Cuffe, Michael, 104
Cullen, Michael, *150*, 166, 167, 168
Cunningham, Alexander, 4, 6
Cunningham, Elizabeth, *145*, 178
Cussen, John, 117, 126, 153, 156, 157

Daunt, George, 16
Davitt, Justice, 124
Dease, Mathew O'Reilly, 46
Dease, Richard, 7, 8, 17, 46
   Dease Memorial, 8
Dease, William, 8, 24, 46
Delaney, Frank, 165
Delaney, Patricia, 156
Dental Hospital, 105, 178
de Valera, Eamonn, 128
diabetes, 115
dietetics, 190
Dimond, Lucy, 135, 174–5
diphtheria, 89
dispensary facilities, 34, 49, 54, 106
   see also pharmacy; skin dispensary
Dr Steevens' Hospital, 4, 38, 94, 117–18, 157, 167, 168, 170, 179, 180
   closure of, 177
Doherty, Evelyn, 178
Doherty, 'Snowy', 178
Dorman, Maureen, 156
Dorset Nourishment Institution, 32
Douglas, Sholto, *147*, *150*, 153, 156, 164, 167

Dow, Robert, 97, 101, 108, *139*
Doyle, P V, 162, 167
Dublin
   architecture of, 3
   the Coombe, 3, 5
   Meath Hospital premises in, 2, 4, 5, 6, 7, 12–13, 14, 15, 17, 20, 46, *57*
   the Liberties, 3, 4, 5, 54
   Long Lane, 54
   Meath Hospital premises in, 15, 16, 17, 18, 20, 25–6, 34, 74
   poverty in, 3–4, 5, 26, 54
   see also Trinity College
Dublin Castle hospital, 93
Dublin County Council Central TB Dispensary, 91, 95
Dublin Hospital Reports, 11
Dublin Institute of Technology, 180
Dublin Journal of Medical Science, 23
Dublin Metropolitan Technical School for Nurses, 175–6
Duffy, Justice Gavan, 125
Duffy, Mary, 178
Duggan, R J, 99
Dwyer, Conway, 81, 82, 83, 87
dysentery, 37

Easter Rising (1916), 94, 95–6
Edge, Heather, 180
Egan, Thomas, 7, 8, 9, 18
election of medical staff *see* appointments procedure; medical board
Emergency *see* Second World War
endocrinology, 166
   thyroid treatment, 117, 155, 166
endoscopy clinic, 169
Ennis, Mary, 6
epidemic wards *see* fever epidemics
erythema nodosum, 106
examinations *see* qualifications

Fallon, Maureen, *149*, 178
famine years, 2, 34
Federated Dublin Voluntary Hospitals, 160, 164, 165
   central council, 157, 158, 191
   see also amalgamation
fever epidemics, 2, 18, 32, 34, 36–7
   separate building for, 37, 50, 54, *58*, 74
   straight waistcoats for, 31
   temporary huts for, 18, 30–1, 32, 33, 34, 36, 49, 54, 74
   closure of, 33
   upper floor beds for, 29, 34, 35, 36, 37, 49, 172
   see also conditions in hospital; food for patients; individual

diseases; Robert Graves; west wing
finances of Meath Hospital, 13, 25, 29, 34, 56, 113–14, 153, 159
   expenditure, 26, 27, 45, 48, 73, 115, 153, 166
   income, 11, 13, 26, 36, 42, 56, 104
   from hospitals' sweepstakes, 105–6
   see also fund-raising; salaries
Fine, Adrian, 159
First World War, 80, 93–4, 103, 174
   temporary hospital staff in, 93
   see also west wing
Fitzpatrick, David, 167
Fitzpatrick, John I, 113, 121, 124–5, *150*, 167, 177
Fleming, Geoffrey, 93
food for patients, 34
   diet of fever patients, 32, 33
   flummery, 31, 33
   hospital kitchen, 153
Foot, Arthur Wynne, 49, 51, 52, 55, 80
Forde, Eileen, 178
fund-raising for Meath Hospital, 11, 45, *69*, 76, 100, 113, *141*, 165, 171, 179, 181–2
   appeals to Grand Jury (1812), 11–12, 13, 17
   bequests, 56, *69*
   Ann Hughes, 74, 75, 76
   Barber bequest, 52–3
   Chester Beatty, 46
   'Jacky' Barrett, 28
   John Bury, 53, 73
   Charitable Sermons, 171
   donations and subscriptions, *72*, 98–9, 153, 156, 166
   Alfred Chester Beatty, 162
   Louis Cohen, 166
   Matt Gallagher, 162
   P V Doyle, 162
   Thomas Pleasants, 14–16
   general appeal (1813), 13–14, 17
   Meath Hospital Guild, 181
   state grants, 34, 166
   subscriptions for wards
   Collis wards, 38–9
   Smyly ward, 45–6, *58*, 106, 160, 162, 173, 181
   see also hospitals' sweepstakes; Hospital Sunday Fund; Irish Stone Foundation; ladies' committee
Furlong, Sydney V, 104, 111, 122, 156, 158, 168

Gage, Winifred, 113, *143*, 175, 175, 176–7
Galashan, James, 103

INDEX

Galbraith, Walter, 128
Gallagher, Matt, 162
Galvin, Hugh J (Joe), *150*, 160, 164
Garland, Eileen, 176
gastroenterology unit (Federated Hospitals), 164, 190
Gatenby, Peter, *147*, 156, 158, 166, 168
   professorial unit, 158–9
General Medical Council, 155
General Nursing Council of Ireland, 175, 176
genito-urinary department, 24, 112, 118, 132, 167
   genito-urinary unit, 116, 118, 133, 134–5, *145*, *149*, 153–4, 157, 165, 167, 178
   first lithotripter, 165
   see also T J D Lane
Gibson, William, 21, 24
Gogarty, Oliver, 83, 86, 87, 94, 98, 103, 106, 107, 122, 123, *140*, 182
   literary career of, 87, 110, 111
   resignation of, 110–11
governors and governesses of Meath Hospital, 55
   see also joint committee
Graham, Ian, 169
Graham, Robert M, 162, 165
Grainger, Ronald, 167
Grand Jury of the county of Dublin, 11, 13, 17, 56–7
Graves, Robert J, 1–2, 18, 21–2, 23, 24, 25, 34, 38, 51, 52, *64*, 65, 182
   appointment of, 18–19
   and fever cases, 29–30, 31, 32, 37
   resignation of, 37
   teaching of, 22, 23, 24, 35, 186
   writings of, 22
Gregory, William, 29
Griffin, John, 157
Griffith, Arthur, 98
grounds of Meath Hospital, 51, 54, 190
Guinness, Arthur (second), 7, 15, 16, 21, 38
Guinness, Benjamin, 7
Guinness, Henry E, 121, 131, 162
gynaecological services, 80, 94, 167

haematology department (Federated Hospitals), 164, 167
Hardwicke Hospital, 29
Harkan, Patrick, 17, 18, 19
Hart, Michelle, 180
Hawkins Street Theatre, 20
Hawkshaw, Henry, 4, 6
Health Act (1953), 136
Health Authority Act (1960), 157

Henry, R F J (Jack), 115, 117, 132
Hepburn, William Joseph, 55, *68*, 73, 83, 172, 183
Hernon, P J, 132
Hewson, Thomas, 10, 17, 20, 25, 38
Hickey, Dr, 126
Hoey, Angela, *151*, 178
Hoey, Elizabeth, 178
Hogan, John, 52
Holles Street Hospital, 105
Home Rule, 76
hospital board *see* joint committee
Hospital for the Incurables, 5
Hospitals' Commission, 115, 179
Hospitals' Federation and Amalgamation Act (1961), 157
hospitals' sweepstakes, 107
   illegal, 99, 113
   legal, 100, 105–6, 115
Hospitals' Trust Fund, 115
Hospital Sunday Fund, 171
Hourigan, Ellen T, 176
Hourihane, Dermot, 160
House of Industry Hospitals, 11
Howard, Palmer, 23
Hudson, Alfred, 38, 41, 43, 52
Huggins, Charles, 153
Hurley, Gerard, *150*, 164, 165, 167
Hurley, Jeremiah, 126–7, 131–2
Hutchinson, Francis, 6
Huxley, Margaret, 175
hypodermic treatment, 39

influenza epidemic (1918), 95
Institution for Sick Children, 102
instruction
intensive care facilities, 46, 160, 162, 190
interdenominational ethos of Meath, 55, 189–90
Irish Health Services Management Institute, 124
Irish Republican Army (IRA), 98
Irish Stone Foundation, 165, 167, 182
Irish Times, 123, 172
   controversy about appointment of P C Smyly, 40–4

Jacob, George Newson, 82, 83
Jervis Street Hospital, 47, 100, 105, 166, 175
Jessop, W J E, 104, 113, 114, 126, 127, 131, *142*, *147*, 153, 155, 156, 180
John Hopkins, Baltimore, 106
Johnson, Stafford, 122
joint committee, 119, 157
   composition of, 76, 77, 78, 121, 125, 130–1
   duties of, 73–4, 132
   Mackey joint committee,

120–2, 126–7, 131, 132, 133
   legal action against, 124, 125–6
   removed from office, 130
   meeting of April 1949, 119–20, 121, 122, 124, 189
   see also Catholics
Jones, Mrs E (matron), 51, 171, 172

Kane, Mary, 179–80
Kane, Robert, 23
Kealy, Revd William, 120, 121, 125
Keating, Albert (Bert), 92, 108, *146*, *147*
Kennedy, David, 191
Kenny, Michael, 76
Keogh, Brian, *150*, 166, 168
Kidd, F W, 80, 94
King's College Hospital, London, 23, 168
Kingston, David, 191
Kirkwood, Major, 120, 121, 125
Knights of St Columbanus, 2, 122, 123, 127–8

laboratory facilities, 50
   haematology services, 160
ladies' committee, 121–2, 135, 159, 161–2, 181
lady superintendents *see* matrons
Lahiff, Mary, 179, 180
Lambert, Mena, 178
Lane, David, *146*, 155, 156, 160, 164
Lane, T J D (Tom), 2, 100–1, 103, 111, 115, 116–17, 120, 121, 122, 123, 124, 126, 127, 131, 132, 134, *139*, *145*, 148, 156, 160–1, 177
   death of, 160
   inaugural address, 115
   pathology, 100, 101
   radiology, 100, 101–2, 108, 115, 176
   urological surgery, 100, 103, 108, 115, 161
   and genito-urinary department, 2, 116–17, 127–9, 134–5, 153, 155, 178, 190
   prostate surgery, 108, 115, 116, 134–5
   see also genito-urinary department
Lane, Victor, 132, *144*, *150*, 153, 156, 160, 163, 167
Lane Joynt, Richard, *72*, 78, 80, 81, 82, 83, 86, 87, 91, 116, *137*, *138*, *140*, 168
   death of, 103
   disabled work, 103
La Touche, Peter, 7, 15

laundry, 54, 76
Leared, Arthur, 23
Leeper, Richard, 107, 109–10
Lees, Cathcart, 37, 38
Leinster, Duke of, 13
Lennon, Edward Emmanuel, 55, 70, 74, 75, 76, 81, 82, 83, 84, 86, 87, 93, 99, 103, 106, 107, 108–9, 116, *137*, *138*, *140*, 155, 179
　death of, 113
Lennon, Emmeline, 155
library at Meath Hospital, 156, 180
lighting in hospital
　candle-light, 34
　gas lighting, 34, 76
Local Government Act (1898), 76–7, 84, 119
Lock Hospital, 8
Lyons, Ellinor, 55, *70*, *72*, 102, *138*, 172, 173, 175
Lyons, Nora, 178
Lyons, Tenison, 55, *72*, 91, 103, *138*
Lytle, Sister Margaret, 94, 97, *143*, 175, 176

Macartney, Professor, 21
Macauley, Charles H, 127
McBride, David, 4, 6, *63*
McCann, John, 131
MacCarthy, Joan, 156
McCarthy, Margaret, *151*, 180
McDermott, T E D, 167, 169
McDonald, George, 159
MacDonnell, Robert, 43
Macdougald, T J, 157
McElwain, John, 167
McGloughlin, S, 121
McGrath, Professor John, 114
McGrath, Michael, 153, 156
McHugh, Barry, 165
McInerney, Ann, 165
McInerney, David, *150*, 167
McKenzie, Stephen, 127
Mackey, Herbert O, 109, 114, 120, 121, 122
　see also joint committee
McLaughlin, Sister Violet, 102
Macnamara, Rawdon I, 20, 25, 39, 46, 172
Macnamara, Rawdon II, 43, 46, 55, *67*, 74, 172, 183
McQuaid, John Charles, 127
Magee, Ann, 113, 131, *148*, 176, 177, 178
Mahoney, James, 76
Maloney, Eileen 'Dan', 179
MANCH group, 164, 178, 191
Mason, Henry W, 91, 93, 100, 101
Masterson, Lottie, 165
Mater Hospital, 79, 167, 175, 179
Matheson, Charles L, KC, 81, 84

matrons, 175–8
　duties of, 35, 50
　residential quarters of, 53
Maxwell, Constantia, 3, 32, 94
Maxwell, Euphan, 94
Mayne, Brian, 117, 118, *147*, *150*, 156, 166, 169, 178
Mayne, Robert, 49
Mayne, Robert St J, 48, 49, 51, 118
Mayo Clinic, 108, 155
measles, 78, 89
Meath Hospital Act (1815), 81, 82
Meath Hospital Act (1951), 126, 127, 129, 130–1, 132, 163, 189
Meath Hospital Dinner, *66*, *71*, 183
Meath Hospital School of Nursing, *151*, 180, 190
medical advisory committee, 163
medical board, 46, 163–4
　composition of, 55, 103
　exclusion of Catholics from, 48
　medical appointments by, 6, 10, 18, 43, 44, 47–8, 73–4, 76–7, 81–7, 130, 132, 185–6, 189
　see also appointments controversy (1910)
Medical Research Council, 159
memorials, 45, 46, *62*, *69*, *146*, 161, 163, 183, 187
　busts presented to hospital, 51, 52, *141*, *142*, 184
　Dease memorial, 46, *59*
　see also fund-raising
Mercer's Hospital, 4, 16, 99, 100, 157
Merrin, A, 100, 101
Montgomery, Douglas, 116, 127, 132–3, *147*, 154, 156, 167
Moore, J N P (Norman), 156
Moore, Sir John William, 52, 55, 74, 76, 79, 81, 82, 84, 87, 90, 99, 103, 104, 105, 106, 109, *137*, *139*, *140*, *142*, 172, 187
Moore, Norman, 108
Moore, William Daniel, 52
Morck, Peter, 46, 160, 162
Morgan, John, 43
Morgan, Owen, 159
mortuary, 54
Moynan, Miss, 179
Mullaney, Joan, 154, 157, 160
　ophthalmic pathology, 160
Mullen, Josephine, 178
Mulvey, Mrs Mary, 120, 121, 123, 125, 131, *145*, 161–2, 182
Munden, P J, 125
Murphy, Ann, 180
Murphy, Cyril, 101, 103, 108, 113, 121, 127, 131, 135, *139*, *140*, *143*, *146*, 155–6, 174, 179, 180

Murphy, Revd J E H, 155
Murphy, Maureen, *144*, 156, 160
Murray, F D, 124, 127, 157

National Blood Transfusion Service, 165
National Children's Hospital, Harcourt Street, 1, 2, 56, 102, 105, 106, 107, 109, 118, 154, 157, 158, 164, 180, 191
National Council for the Blind, 168
National Haemophilia Centre, 164–5
National Insurance Act (1911), 90
National and Orthopaedic and Children's Hospital, 102
Newell, F T Porter, *70*, 172
Norman, Connolly, 159
nurses (and wardsmaids), 50, 95, 112, *142*, 170
　accommodation for, *59*, 116, 133, 134, 173–4, 177
　1890s staff, 55
　pension scheme for, 174
　Red Cross Sisters *see* Red Cross Order of Nursing Sisters
　religious orders, 170, 175, 179
　training of probationers, 55–6, 78, 102, *151*, 170–1, 172, 174, 175–6, 177–8, 180, 190
　tutor sisters, 178
　urological, 108, 117, *149*, 178
　see also Dublin Metropolitan Technical School for Nurses; General Nursing Council of Ireland; Past Nurses' Union; Queen Alexandra's Nursing Service; salaries
Nurses Registration Act (1919), 175
nursing committee, 173, 174, 175, 176

O'Brien, Brendan, 108, 112, 114, *144*, 156, 166
O'Brien, Dermod, 114
O'Brien, Kitty, 182
O'Brien, Mabel Emmeline, 114
O'Brien, William Smith, 114
occupational therapy, 190
O'Connor, Patrick J, 121, 131, 162
O'Connor, Alderman William, 108
O'Dwyer, Elizabeth, *148*, *151*, 177–8
O'Flynn, Dermot, 128, 132, *144*, *150*, 153, 156, 160, 161, 165, 167, 169
O'Hanlon, Roderick H (Rory), 126–7, 132, *147*, 156, 166, 168
O'Higgins, Kevin, 100
O'Higgins, T F, *145*, 153
O'Kelly, Dermot, 126

INDEX

O'Leary, Jack, 108, 117
Oliver, Silvia Deane, 104, 126, 155, 156
O'Loughran, Frank, *150*, 168
O'Meara, Professor, 114, 153
O'Morain, Colm, *150*, 169
O'Neill, Thomas, 126–7
operations, 37
  facilities for, 14, *72*, 76, 106
  in wards, 13
'Orange outrage', 20–1
O'Reilly, Helen, 179
O'Reilly, J J, 76
O'Riordan, Jack, 165
Ormsby, Sir Lambert, 8, 49, 51, 55, *68*, 73, 76, 81, 82, 83, 84, 86–7, 171, 172
  affair of the chauffeur's boots, 102–3
  death of, 94, 102
  Medical History of the Meath Hospital, 1, 38, 46, 55, 73, 102
orthopaedic services, 102, 104, 164, 165, 167, 177, 179, 190
O'Shaughnessy, I L, 98
Osler, William, 23
O'Sullivan, Kevin P, *150*, 157
Our Lady's Hospice for the Dying, Harold's Cross, 90
out-patient facilities, 13, 34, 49, 153, 162
Owens, Gerald, *150*, 165

parathyroid surgery, 154, 155, 168
Past Nurses' Union, 179
pathology department, 76, 98, 114, 153, 160
Patteson, Robert Glasgow, 80
Pearson, William, 83, 84, 86
Peel, Robert, 17
Pegum, John Michael, *150*, 164, 167
Penal Laws, 5
Penrose, Francis, 55, *137*
peritoneoscope, use of, 133
pharmacy, 162
Phelan, John, 157
physiotherapy department, 159, 179, 190
Pleasants, Thomas, 14–16, 17
  built Stove Tenter House, Cork Street, 16
  donations of, 16
Plunkett, Horace, MP, 77, 98
Pole, W W, 11
Population Act (1814), 3
Porter, Sir George Hornidge, 45, 47, 55, *67*, 74, 102, 109–10, 172
Porter, William Henry, 20, 24, 40, 41, 43, 45, 55, *64*, 110
porters, 50, 92
  Porter's Lodge, 54

Portobello House, 176
post-mortem room, 54
Powerscourt, Lord, 13, 14, 74, 90
Pringle, Harold, 115, 135, 153
prizes *see* qualifications
professorial unit *see* Peter Gatenby
psychiatry department, 107–8
Public Charitable Hospitals Act (1930), 99
Purser, Francis C, 81

qualifications
  examinations and prizes, 35, 163, 175, 187–8
  Hudson Scholarship, 38
  Lucy Dimond award, 180
  Nora Lyons award, 180
  Sir John Moore Prize, 187–8
  Smyth Memorial Scholarship, 187
Queen Alexandra's Nursing Service, 176
Queen's College, Belfast, 23
Queen's College, Cork, 23
Quilty, Fergus, 156
Quin, James S, 121, 131, *143*, *147*, 156, 168

radiology department, *72*, 78, 80, 91, 98, 100, 106, 108, 115, 135, 153–4, 167, 176, 190
computerised tomography, 167, 190
see also T J D Lane
rebellion of 1798, 11
Red Cross Order of Nursing Sisters, 55, *70*, 78
  Sisters' House, 56, 102
  Training School, 102, *138*, 172, 173, 178
registrar or register *see* secretary
regulations for patients (1890s), 56
relapsing fever, 37
renal unit, 166, 168, 190
  kidney disease, 165–6
  transplants, 166
  see also genito-urinary unit
research *see* clinical research
Rest for the Dying, Camden Row, 90
rheumatic fever, 106
Richards, Solomon, 7, 8, 10, 17, 25
Richmond Hospital, 47, 49, 81, 87, 117–18, 122, 176
Ricketts, Howard, 29
Road Act, 11
Robinson, Cecil, 93, 94, 104, 107, 120, 125, *143*, *147*, 162, 183
Robinson, Derek, *147*, *150*, 154, 156, 183
Rockefeller Foundation, 99, 104
Rogers, R A P, 187

Roney, Cusack, 7, 8, 20, 25, 34, 45
Roney, Patrick Cusack, 7, 8
Roney, Thomas, 8, 17, 20, 25, 38
Rooney, Eamon, TD, 121, 131
Ross, Charles Homan G, 81, 82, 84, 87
Rotunda (Lying-In) Hospital, 4, 39, 106, 107
Royal Academy of Medicine, 109
Royal City of Dublin Hospital, 79, 99, 171, 176
Royal Dublin Society, 16
Royal Hospital for Incurables, 90
Royal Hospital Kilmainham, 79
Royal Military Hospital, Montpelier Hill, 93
Royal National Hospital for Consumption, Newcastle, 90
Royal Southern Hospital, Liverpool, 171
Royal Victoria Eye and Ear Hospital, 160, 175
Russell, Angela, 120, 121, 131, 181
Ryan, James, 134
Rynd, Francis, 24, 39, 41, 43, 44, 66

St Catherine's Annexe, 135, 162
St James's Hospital, 164, 168, 180
  haemophilia unit, 167
St John's Ambulance Brigade, 133, 154
St Kevin's Hospital, 134
St Mary's Annexe, 112, 162, 166
St Patrick's Hospital, 5, 107, 109
St Vincent's Hospital, 79, 114, 175
salaries of hospital staff, 27, 31, 35, 50, 73, 101, 104
  apothecary, 91, 103
  cook, 50
  dividend from teaching fees, 186–7
  doctors, 31, 32, 34
  ISA fund (doctors' pool), 136–7, 163
  matron, 173
  nurses, 73, 91, 115, 170, 173, 174, 176
  probationers, 176
  surgeons and physicians, 73, 77
scarlatina, 89
School of Physic, 23
Second World War, 112, 176, 177, 178
secretary, 50, 55
  duties of, 27, 35, 50
  residential quarters of, 50, 53
Shanahan, Edward, 126
Sharma, Deen, 159
Sheridan, Desmond, 159
Sheridan, Eileen, *149*, 176

Sick and Indigent Roomkeepers'
  Society, 5
Sir Patrick Dun's Hospital, 80, 95,
  97, 99, 100, 105, 116, 133,
  157, 164, 168, 174, 176
skin dispensary, 108–9, 114
Slattery, Richard Vincent, 86
smallpox, 49–50, 89
Smyly, Josiah, 24, 39, 40, 43, 44,
  45, 46, *66*
Smyly, Philip Crampton, 24, 39,
  55, *67*, *72*, 114, 171, 172
  appointment of, 40–4
Smyly, Vivienne, 39
Smyly, William Josiah, 39
Smyth, Joseph, 187
social work, 179–80, 190
Somerville-Large, W Collis, 104
Spilsbury, Bernard, 88
standing committee *see* joint
  committee
statistics at Meath Hospital, 37
  number of beds, 26, 37,
    171–2, 173
  for fever patients, 32, 35, 37
  in GU unit, 153
  for traumatic orthopaedics,
    167
  number of patients
    in 1892, 56
  fever patients, 49, 78, 89
  haemodialysis patients, 166
  influenza patients, 95
  wounded soldiers, 94
  of X-rays, 115
Steen, Robert E, 109, 118, 123,
  158
Stephens, E Brandon, 116, 126,
  132, *147*, *150*, 154, 155, 156,
  168–9
Stewart, Professor F S, 153
Stokes, Adrian, 94, 95, 104
Stokes, Barbara, 125, 168
Stokes, Frederick, 47–8
Stokes, Henry, 82, 86–7, 93, 94,
  95, 96, 99, 103, 104, 107, 109,
  114, 115, 116, 117, 120, 121,
  123, 126, 132, 134, *139*, *140*,
  142, *143*, 154–5, 156, 162,
  168, 179, 183
  resigned, 154
Stokes, Whitley, 18, 20, 21
Stokes, William Snr, 2, 11, 18, 21,
  22, 23, 24, 25, 34, 37, 38, 41,
  43, 46–7, 51, *64*, *65*, 80, 104,
  109, 114, 182, 184
  death of, 52
  *An Introduction to the Use of the
    Stethoscope*, 22
  reply to Lord Lieutenant, 34–5
  Stokes–Adams Syndrome, 22
  teaching of, 22, 23, 24, 51, 186
  writings of, 22
Stokes, Sir William, 46–7, 49, 52,
  55, *68*, *72*, 80

Stronge, James Whitelaw, 47
students at the Meath, 23, 35, *71*,
  93, 136, *147*
  bedside teaching method, 21,
    186, 187
  foreign, 42
  numbers of, 36
  residence, 136, 187
  surgery, instruction in, 24, 35,
    42
  apprenticeship method, 25,
    65, 186
  teaching fees for, 73, 186–7
  unruly behaviour of, *141*,
    183–4
  see also teaching reputation of
    the Meath
sweepstakes *see* hospital
  sweepstakes
Swift Jonathan, 15
  Dean's Vineyard, 15, 37, *57*
Synge, V M, 114

Tallaght Hospital *see*
  amalgamation
Taylor, Mervyn, 159
Taylor, Robert, 108
Taylor, William, 80, 81, 82–3, 87,
  93, 94, 97, 100, *137*, *138*, 181
Teac Ultain Hospital, 105
teaching reputation of the Meath,
  42, 186–8
  see also names of individual
    medical staff; students at the
    Meath
telephones in hospital, 76
Temperley, Ian, *147*, 158, 159,
  160, 164, 166, 167
Terry, Joseph, 5
Thompson, Gershon, 108
Thorndyke Laboratory, Boston,
  166
Thornhill, John, 167
Todd, Robert Bentley, 23
Todderick, Dr, 10
Trinity College, Dublin, 18, 25,
  28, 41, 47, 52, 80, 113, 133,
  153, 155, 158, 160, 161, 164,
  165, 168, 178, 180
  Trinity department of clinical
    medicine, 164
  Trinity professorial unit, 159
Trousseau, Armand, 23
tuberculosis
  pulmonary, 89–91, 107, 114
  1940 epidemic, 112–13
  urological, 135
  see also Dublin County Council
    Central TB Dispensary
typhus fever, 8, 18, 29–30, 31, 37,
  74, 78, 89

United Irishmen, 18
urological surgery *see* genito-
  urinary department; T J D Lane

vaccination, 49
venereal disease, 94–5
  see also Lock Hospital
Vesey, Captain, 76
Victoria, Queen
  visit to Dublin, 78–9, *137*

Wade, Georgina, 180
Wall, Mary C, 97, *139*, 175, 176
Walpole, George, 172
Waterfield, W H, 183
Waterloo, battle of, 16
Watson, John
  *Gentleman's and Citizen's
    Almanac*, 4
Webb, Marcus, 159
Weir, Donald, *147*, 158, 159, 164
Wellcome Trust, 165
Wellesley, Marquis, 20
west wing, 54, 76, 78, 89, 106,
  107, 112, 113, 133, 134, 168,
  173, 175, 176, 177
  accommodation of wounded in
    First World War, 94
  construction of, 74–5
  coronoary care
    accommodation, 164
  non-infectious cases, 104
  tuberculosis wards, 91
  see also St Mary's Annexe
Wharton, James Henry, 41, 43,
  46, 48, 172
White, E Bantry, 93
White, Robert Persse, 43, 47, 49,
  73
Whitelaw, Reverend James, 4
Widdess, Professor J D H, 5, 20,
  24
Wilde, Sir William, 23
Wilson, C H (Bertie), 153, 156
Women's National Health
  Association of Ireland, 90
Woods, Sir Robert, 122
woollen weavers, 5, 16
works committee, 133, 134, 162,
  168, 177

X-rays *see* radiology

Young, Augusta, 109, 114, 157